CROP PROTECTION MONOGRAPHS

CROP PROTECTION MONOGRAPHS

Already published volumes

J. Palti, R. Ausher (Eds.)
Advisory Work in Crop Pest and Disease Management

H. Waibel
The Economics of Integrated Pest Control in Irrigated Rice

Further volumes in preparation

D. P. Singh
Breeding for Resistance to Diseases and Insect Pests

I. Wahl and G. Fischbeck
Role of Wild Relatives in Control of Crop Diseases,
Development of Their Epidemics, and Evolution of Pathogens

Hermann Waibel

The Economics of Integrated Pest Control in Irrigated Rice

A Case Study from the Philippines

With 19 Figures and 81 Tables

Springer-Verlag
Berlin Heidelberg New York
London Paris Tokyo

Dr. HERMANN WAIBEL
Thai German Plant Protection Programme (TGPPP)
P.O. Box 9-100
Bangkok 10900, Thailand
and
7077 Alfdorf-Stixenhof
West Germany

ISBN-13:978-3-642-71321-7 e-ISBN-13:978-3-642-71319-4
DOI: 10.1007/978-3-642-71319-4

Softcover reprint of the hardcover 1st edition 1986

2131/3130-543210

To my father
and to my late mother
for their motivation
and to my family
Mila Linda,
Matthias and Christoph
for their patience

Acknowledgements

I am indebted to Prof. Dr. J. Zeddies, Professor of Agricultural Economics at the University of Hohenheim, and to Prof. Dr. J. Kranz, Professor for Phytopathology and Applied Entomology of Tropical Agriculture for their continuous help and advice.

During the field study in the Philippines I was supported by numerous persons and institutions. To all of them I am indebted for their personal help and support. I am particularly thankful to Mr. J. Sumangil, the Chief of the Crop Protection Division of the Bureau of Plant Industry. Mr. Mauricio Bayoneta, Acting Chief of the regional Crop Protection Center in Iloilo deserves my special gratitude for his unselfish support of my work. I should also like to thank the other members of the regional Crop Protection Centers from Iloilo as well as the staff from Pili and Maligaya for the execution of the cooperatively planned field trials. I am also indebted to the German staff of the Philippine German Crop Protection Programme, especially to Mr. R. Kaske, GTZ, for his outstanding and continuous support.

In the technical preparation of this work I was assisted by Mrs. Agnes Bardoll, Mrs. Barbara Obert-Loercher, Mrs. Karin Althaus from the Institute of Agricultural Economics, University of Hohenheim, Mrs. C. Mode translated the text from the German and Miss Susanne Koch prepared the figures.

The study was financed in part by the Deutsche Gesellschaft für Technische Zusammenarbeit (GTZ), the German Academic Exchange Service (DAAD) and the University of Hohenheim. These institutions deserve my special thanks.

Last but not least I want to thank for their patience and cooperation the Filipino rice farmers who were willing to provide their fields for on-farm trials and to answer questions during interviews. As with many other researches, it is also likely in this case that the farmers' contributions to the solutions found in this study were larger than the researcher's.

The greatest thanks, however, has to go to my wife, Mila Linda, who had the greatest, although invisible effect on the successful completion of this book and who had to bear the major share of the "social costs" involved.

HERMANN WAIBEL

Contents

List of Abbreviations

AC	Armyworm/cutworm
BAEX	Bureau of Agricultural Extension
BPH	Brown Planthopper
BPI	Bureau of Plant Industry
BRES	Bicol Rice Experiment Station
CI	Confidence Interval
CW	Caseworm
DAS	Days after seeding
DAT	Days after transplanting
DS	Dry season
GLH	Green Leafhopper
HYV	High Yielding Varieties
IRRI	International Rice Research Institute
LF	Leaffolder
MRRTC	Maligaya Rice Research and Training Centre
n.a.	not available
OLT	Operation Land Transfer
P	Peso, 1 Peso = 0,28 DM
RB	Rice Bug
RS	Rainy season
SEWS	Surveillance and Early Warning System
VRES	Visayas Rice Experiment Station
WM	Whorl maggot

Introduction

As a result of the green revolution, the use of yield-increasing inputs such as fertilizer and pesticides became a matter of course in irrigated rice farming in Southeast Asia. Pesticides were applied liberally, both as a guarantee against crop failure and as a means of fully utilizing the existing yield potential of the crops.

However, since outbreaks of pests, such as the brown planthopper (BPH) or the tungro virus, continued to occur despite the application of chemicals, a change of approach began to take place. It is now being realized more and more in Southeast Asia that crop protection problems cannot be resolved solely by the application of chemicals. In the past several years, increasing efforts have therefore been made to introduce, as a first step, supervised crop protection, leading gradually to integrated pest management (Kranz, 1982).

Although the crop protection problems naturally differ in the different developing countries in Southeast Asia, the economic situation prevailing in these countries can nevertheless be regarded as an important common determinant: pesticide imports use up scarce foreign currency and thus compete with other imports essential to development.

For the individual rice farmer, the problem is basically the same: his cash funds are limited and he must carefully weigh whether to use them for purchasing pesticides, fertilizer or certified seed. In view of this constraint, it is becoming necessary to abandon the purely prophylactic, routine calendar spraying and instead, employ critically timed and need-based pesticide applications.

The decision problems involved seem to call for a detailed economic analysis of crop protection.

In the following, a case study will be presented, taken from irrigated rice farming in the Philippines. Three pilot projects have been selected as examples of the introduction of integrated pest management.

Farmers' surveys were conducted in the investigation area to collect socio-economic data and information on production techniques, and field trials were also carried out. The trials concerned were primarily aimed at determining damage coefficients and quantifying the control effectiveness of alternative crop protection strategies. However, since they were conducted under practical conditions on farmers' fields, the results arrived at can serve merely as points of departure. All the data obtained by the author from his own surveys and trials or from other sources, were then used in a comparatively simple, decision-theoretic test model. From the results arrived at by means of this model, inferences can be drawn regarding the optimum crop protection strategies for specific objectives and the benefit to be derived from different stages of improvement of the information basis.

This research attempts to present possible methods, geared to conditions in developing countries, of dealing with problems of this kind. The primary goal was to find a practicable compromise, on the basis of the available data, between the requirements of crop protection and economics.

1 Background Conditions and Current Situation in Irrigated Rice Farming

1.1 General Cropping Conditions

In the Philippines, irrigated rice accounts for only about 35% of the total crop acreage, but makes up over 60% of the country's rice production. Since 1960, due to the zealous efforts of government institutions to set up canal irrigation systems and smaller-scale pumping schemes, the irrigated acreage has more than doubled (Bureau of Agricultural Economics, 1980).

Provision of irrigation water, having eliminated one of the main contraints to the full utilization of the yield potential of the *high yielding varieties* (HYV), it became possible to grow rice during the dry season, i.e. at a time when climatic conditions are much more favourable, so that higher yields can be achieved and crop failures as less likely.

Yield levels which amount to 2.75 tons ha^{-1} (Table 1) are nevertheless still considerably below the economically desirable figure. In this context it must

Table 1. Development of crop area, production and yield per ha in irrigated and non-irrigated rice in the Philippines, 1970–1981

Year	Crop area (1000 ha)			Production (1000 t)			Yield t/ha		
	Irri-gated	Non-irrig.	Total	Irri-gated	Non-irrig.	Total	Irri-gated	Non-irrig.	Total
1970	1345.7	1767.7	3113.4	2716.0	2517.4	5233.4	2.05	1.42	1.68
1971	1470.5	1705.1	3112.6	2930.8	2412.1	5342.9	2.0	1.41	1.72
1972	1332.0	1914.4	3246.4	2617.3	2482.7	5100.0	1.96	1.3	1.57
1973	1241.0	1870.8	3111.8	2343.6	1399.5	4414.6	1.89	1.34	1.42
1974	1493.7	1943.1	3436.8	3015.1	2579.0	5594.1	2.02	1.33	1.63
1975	1411.7	2127.1	3538.8	3033.9	2566.5	5600.4	2.15	1.21	1.60
1976	1494.6	2084.7	3579.3	3369.8	2789.7	6159.5	2.25	1.34	1.72
1977	1489.6	2057.9	3547.5	3494.0	2962.0	6456.0	2.35	1.44	1.82
1978	1514.9	1994.0	3508.9	3933.4	2961.5	6894.9	2.6	1.49	1.97
1979	1472.0	1996.9	3468.9	4051.6	3145.9	7197.5	2.75	1.58	2.08
1980	n.a.	n.a.	3503.1	n.a.	n.a.	7504.4	n.a.	n.a.	2.14
1981[a]	n.a.	n.a.	3459.1	n.a.	n.a.	7722.7	n.a.	n.a.	2.24

[a] n.a. = data not available

Source: Bureau of Agricultural Economics, Ministry of Agriculture, Manila 1980

however, be noted that yields are often underestimated in the official statistics, since the figures given by the farmers tend to be lower than actual outputs. Furthermore, labour is paid in kind, about ⅙th of the crop, which is generally not included. In addition, the size of crop acreage is not indicated exactly, i.e. it is rounded off to the next half or whole hectare. Nevertheless, it can be assumed that actual yields fall about 25% short of the economically optimum level. This is due, inter alia, to economic cause (e.g. shortage of finance) and to the fact that the farmers' attitude is not sufficiently profit-oriented (Herdt and Wickham 1978).

As regards *farm sizes,* the majority of the farms are small holdings, the average acreage in the Philippines being 2.9 ha per farm. The rice farms are even smaller, with acreages of generally 1–2 ha per farm.

Farm sizes must be viewed in close relation to the prevailing conditions of *land tenure.* In 1972, an agrarian reform was introduced covering rice and maize cultivation. Within the scope of the *Operation Land Transfer,* privately owned land exceeding 7 ha per gainfully occupied person was assigned to the former tenants. The reform was applied first of all to the owners of more than 100 ha of land, whilst smaller landlords were left out. For the latter group, a separate mode of procedure was introduced, the so-called *Operation Leasehold,* a change-over from share-cropping to a form of tenancy with a fixed rent. This was generally based on the average yield of the past 3 years and the share-croppers were provided with registered leasehold contracts. However, since only part of the crop acreage was included in this reform program, share cropping still plays a considerable role in the structure of land tenure. By 1977, only about 30% of the 1.42 million ha of land which are to be thus transferred had been assigned to the former share croppers (Philippine Statistical Yearbook 1978). As far as the irrigated areas are concerned, it can be assumed that the leasehold type of *land tenure* still represents the most common type of tenancy.

As regards the natural crop production conditions, there is also a considerable difference in irrigated farming between the rainy and dry periods. During the rainy season, climatic factors greatly affect yield levels, causing considerable annual fluctuation. Thus, during this period, which may last from May to November, depending on geographical location, about 20 typhoons can be expected to occur every year, often producing heavy floods. Furthermore, in the rainy season, the increasingly heavy rainfall produces technical problems resulting in a considerable lengthening of the planting period. As Fig. 1 shows, the period from June to September can be regarded as the main planting time in the rainy season, while only December and January are peak planting periods in the dry season. During the dry season, the period of water availability is relatively shorter than in the rainy season, a fact which is often reflected in the planting method.

In the rainy season, the predominant method of planting consists in raising the rice seedlings in seedbeds and subsequently transplanting them; in the dry season, direct seeding is becoming increasingly common. The length of the planting period in the rainy season has considerable consequences for the crop protection extension work, in particular for the *pest surveillance* programme.

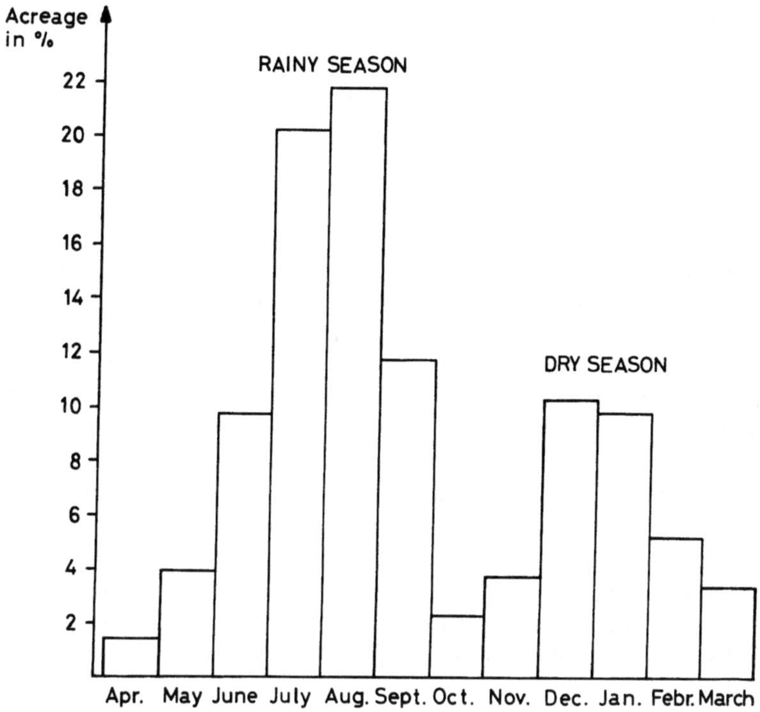

Fig. 1. Distribution of rice planting periods in the Philippines

1.2 Organization of Extension and Crop Protection Services

Following the organizational structure customary in the highly developed agricultural economies, crop protection service and general extension service in the Philippines were also established as two seperate institutions. Since the general extension service was assigned the task of implementing production programs, it was perforce also confronted with crop protection problems. The crop protection service, on the other hand, was generally unable to take any action unless problems arose.

This situation has often resulted in overlapping spheres of responsibility and contradictory recommendations.

When the Ministry of Agriculture was reorganized in 1980, this fact was taken into account and a joint organizational structure was set up from the regional level downwards.

1.2.1 Organizational Structure of the General Extension Service

Since the beginning of 1980, the extension service has been handled directly by the Ministry of Agriculture via its agencies at the regional and provincial level. A great part of the extension work concerns the implementation of a rice production promotion programme. This so-called *Masagana 99 programme*[1] was initiated in 1972 and consists essentially of four elements:

- Provision of non-collateral loans for production inputs,
- Government buying at guaranteed prices,
- Provision of subsidized fertilizer, and
- Provision of extension recommendations for production techniques by trained technicians.

Only the irrigated rice acreage on which two crops are raised per year is covered by this programme. Despite considerable initial success, the programme is today characterized by a very low rate of repayment of the production input loans. Although the rate of repayment still amounted to about 91% in 1973/74, it had dropped to about 35% by 1975/76 (Chandler 1979). As a direct consequence, the loan programme was reduced so that today it only covers about 12% of the originally planned acreage, i.e. about 125000 ha (NFAC 1980).

Loans for land preparation are provided in cash; for yield-increasing/yield securing inputs, such as fertilizer and pesticides, they are in the form of coupons. Since the extension workers have to look after about 100 farmers each, their work is based on a rigid schedule of recommendations (16 steps of rice production). As far as crop protection is concerned, this mode of procedure results almost automatically in the recommendation of prophylactic measures.

1.2.2 Organizational Structure of the Crop Protection Service

Crop protection is one of the main tasks assigned to the Bureau of Plant Industry (BPI). The Crop Protection Division of the BPI is responsible for drawing up national guidelines and preparing crop protection programmes. Implementation of the programmes, on the other hand, is the responsibility of the Crop Protection divisions forming part of the regional agricultural authorities. Corresponding divisions exist at provincial and district levels.

The *Surveillance and Early Warning System (SEWS),* set up in 1974, must be viewed in the light of the large-scale pest calamities which occured in particular during the first half of the 1970's (Table 2). Since such pest outbreaks continued to occur despite the *Masagana 99 programme* with its recommend prophylactic crop protection measures, plans were made to develop a system which would provide information at an early stage regarding the necessary additional control measures to be taken. This system was to be based on systematic and continuous pest monitoring and on provisionally fixed pest population *control thresholds.* Selected farms used as *observation stations* formed the basic elements of this *SEWS*. However, since the *SEWS* had been set up as a separate unit and had not been integrated into the general extension service, additional *extension recommendations* could be issued only when the given thresholds for important pests were surpassed. Furthermore, since the input loans were granted via the general

1 Masagana is a Philippine word signifying "plentiful". The figure *99* represents the production target of 99 cavans ha^{-1} or approx 4.4 t ha^{-1} (1 cavan = 44 kg)

Table 2. Some of the major pest calamities in rice cultivation in the Philippines

Year	Pest	Infested area (ha)	Location
1940	Tungro virus	1000000	Country-wide
1958	Rats	600000	Country-wide
1970/71	Tungro virus	100000	Central region, Bicol
1973/74	Brown planthopper	10000	Central region
1974	Rats	850000	Country-wide
1975	Tungro virus	10000	Country-wide
1976	Brown planthopper	17000	Mindano

Source: Annual reports of the Bureau of Plant Industry, Manila, different issues

extension service, the recommendations made by the warning service often produced no effect because of the farmers' shortage of cash.

Since the above-mentioned reorganization of the Ministry of Agriculture (1980), responsibility for warning service activities has been in the hands of the Crop Protection Departments at the regional Agricultural Authorities. The *regional Crop Protection Centers,* on the other hand, are responsible for the technical supervision and the training of the crop protection extension staff.

The *crop protection recommendations* of the extension service are drawn up annually by a so-called Interagency Committee consisting of representatives of the Ministry of Agriculture, the International Rice Research Institute (IRRI) and the National Crop Protection Center. These *control recommendations* indicate the prophylactic measures to be taken and also include instructions regarding the use of *control thresholds*. However, the thresholds given are very approximate rule-of-thumb values and are not based on especially designed research.

1.3 The Major Pests of Rice

Apart from weather-dependent factors, *crop losses* in rice are determined by insects, diseases, weeds and, in particular, rats. Actual data on the losses attributable to the individual causes are sparse. Although subjective *crop loss assessments* are made within the scope of the general crop statistics of the Bureau of Agricultural Economics, these figures do not seem to be very reliable.

To date, the importance of the individual causes of damage has been measured only in terms of the size of infested areas. The basic data for these assessments are elaborated by the BP with the assistance of its extension workers at village level.

According to official statistics, a greater proportion of crop acreage is infested by insect pests than by diseases, as shown in Table 3. Studying this table, a definite downtrend can also be noted in infected/infested areas over the past several years as a result of the increased use of resistant varieties. Rats and weeds are not listed in the table. For rats, a new kind of survey method was

Table 3. Rice area (ha) infested/infected by insect pests and diseases 1967–1979[a]

	Insect pests (ha)				Insect pests total	Diseases (ha)			Diseases total	Diseases and pests	Infested area as percent of total crop acreage
	Plant hoppers	Stem borer	Leaf pests	Others[b]		Virus	Fungal	Bacterial			
1967	2880	38420	15090	16487	72877	301	3351	9975	13627	86504	
1968	n.a.	n.a.	n.a.	n.a.	n.a.	n.a.	n.a.	n.a.	n.a.	n.a.	
1969	3528	51081	19773	15491	89873	30	704	45	779	90652	2.72
1970	6222	44991	30263	20344	101820	2603	901	2415	5919	107739	3.46
1971	n.a.	n.a.	n.a.	n.a.	n.a.	100000[c]	n.a.	n.a.	–	–	–
1972	16785	7506	62232	24998	111521	6627	14862	3661	25150	136671	7.0
1973	31587	57615	39716	23877	152795	4200	3112	2084	9396	16191	8.7
1974	105038	79052	54581	120143	358814	2624	15786	3960	22370	381184	18.5
1975	n.a.	n.a.	n.a.	n.a.	288019	n.a.	n.a.	n.a.	22508	310527	14.6
1976	30000	105000	70000	27000	232000	11218	4475	4500	20193	252193	11.7
1977	21000	63000	82000	25000	191000	7343	2570	2000	11913	202913	9.5
1978	6000	31000	72000	9000	118000	9964	10270	15414	35648	153648	7.3
1979	7226	1899	4461	660	14246	2498	4572	12563	19633	33879	1.6

[a] Approximately 50% of the area is supervised
[b] Rice bug and locusts
[c] Estimate
n.a. = data not available
Source: Annual Reports, Bureau of Plant Industry, Manila

Table 4. Important weed species in rice in the Philippines (DeDatta 1976)

Species	Irrigated rice	Upland rice
Grasses	*Echinocloa crusgali*	*Echinocloa colona* *Digitaria sanguinalis* *Eleusine indica* *Dactyloctenium aegyptum*
Sedges	*Cyperus difformis* *Cyperus iria* *Fimbristylis literalis*	*Cyperus rotundus*
Broad leaved weeds	*Monochoria vaginalis* *Sphenochlea zeylenica*	*Impomea triloba* *Portulaca olerazea* *Commelina benghalensis*

Table 5. Crop loss due to rats, in %

Year	Minimum	Maximum	National average
1975	0.47	10.57	4.63
1976	0.15	1.25	0.39
1977	0.09	0.91	0.45
1978	0.01	12.09	0.71
1979	0.08	3.84	0.40

Source: Annual Reports, Bureau of Plant Industry, Manila, 1975–1979

adopted in 1976, the criterium now used being infestation intensity, i.e. percentage of damaged tillers as opposed to size of infested area, the measure of assessment formerly used. For weeds, the figures available are very inadequate. Table 3 lists the major pests and diseases in irrigated rice farming in the Philippines, Table 4 shows the most important weed species.

Although more than 100 insect species are known to infest rice, only about 20 of these can be regarded as important pests (Pathak 1977b). In Philippine crop protection, only eight of these are considered to be major pests.

Going by the control measures implemented, virus diseases rank first among the great number of existing rice diseases. Although on the average the losses caused by fungi may well be higher, minimal use has been made of fungicides to date, even in the areas with intensive rice cultivation. This is mainly due to the high costs involved.

The same applies to bacterial diseases. Although bactericides are available, they are hardly ever employed in practice, resistant varieties being the most important control measure actually used.

Although *varietal resistance* represents the main component of crop protection for all diseases, the in-bred resistance genes cease to be effective against the predominant pest races after a certain time, i.e. a so-called *resistance break-down* occurs in the long term. In the case of virus diseases, vector control is therefore also of considerable importance as a means of preventing transmission of the viruses by planthoppers during the particularly susceptible early growth stage of the rice plant.

Apart from some of the pathogens listed in Table 3, rats are the major pests in rice, damaging the rice plant at all growth stages. In 1958, damage caused by rats amounted to 40 million Pesos, excluding post-harvest losses (Alfonso and Sumangil 1977). Up to the beginning of the 1970's, rats were therefore a major output reducing factor (Table 5).

Control measures include improved sanitation, cultural practices and physical measures, as well as the coordinated use of acute and chronic poisons.

2 Assessment of Crop Loss Due to Pests in Irrigated Rice Farming

For preparing a valid economic analysis of crop protection measures, quantification of the losses due to pests is an indispensable precondition.

In rice farming, systematic crop loss surveys conducted at regular intervals were started only a few years ago and have to date been publicized only fragmentarily. Most of the available data are consequently either rough estimates or else trial results, which often do not permit a distinction to be made between pest influence on yields and the effects produced by the chemicals applied.

Apart from various global *loss assessments* (Cramer 1967; Barr et al. 1981), the first field study for determining the losses caused by the stem borer was carried out in the Philippines in 1972 (Gomez and Bernardo 1974). This was followed by glasshouse studies, in which the losses caused by specific pests were simulated by means of artificially produced damage symptoms, and by cage trials with controlled pest infestation levels (Dyck 1978). It was not until recently that a *methodology of crop loss assessment* related to the different plant growth stages was developed by Litsinger et al. (1980).

According to Cramer's study (1967) on crop losses due to pests – probably the most often quoted work on this subject – overall *crop losses* in rice cultivation in Asia (excluding the People's Republic of China) amount to 51.5%, whereby 31.5% are due to pests, 9.3% to diseases and 10.7% to weeds. For the Philippines, Cramer directly uses the figures computed mainly from Indian data, i.e. losses of 36% due to pests, 10% due to diseases and 11% due to weeds (Cramer 1967). Pathak and Dyck (1973) arrived at similar loss figures: from a total of 67 insecticide trials (1964–1967), they computed a mean difference of 20–25% between treated and untreated plots. According to a survey covering several countries (Barr et al. 1981), the stem borer alone can be assumed to cause losses of 6.6% in the Philippines in normal years.

2.1 Assessment of Crop Loss in the Philippines

For determining *crop loss* in rice, there are various types of field trials originally designed for other purposes. First of all, there are the official pesticide evaluation trials conducted at regular intervals at various experimental stations. In addition, there are the IRRI on-farm trials. These include trials carried out to determine the extent to which further yield increases can be achieved by means of higher rates of insecticide application. Furthermore, data are available from trials conducted according to the Litsinger method. Since in these trials treatment variants with different rates of application were compared with untreated control plots, they are basically the most suitable trials for *crop loss assessments.* In all the above trial series, *crop loss* is derived from the yield difference between the most heavily infested and the least infested plots.

In collaboration with the *Regional Crop Protection Centers,* the author also conducted some trials of his own following the Litsinger concept (Chap. 2.1.3). In these trials, the commonly used crop protection practices of the local farmers were included as trial variants.

2.1.1 Official Pesticide Evaluation Trials

A great number of trials are being carried out to test the biological effectiveness of different insecticides. To obtain adequate infestation levels, the IR-22 variety is generally used in the Philippines, since it is regarded as being susceptible to all major pests. On small plots (20–30 m^2), the level of attack by major pests is assessed and grain yields are determined. *Crop losses* are then computed from the difference between the respective maximum yield and the untreated control plot. For assessing *crop losses,* it is assumed that the maximum yield corresponds to the *potential yield,* i.e. to the yield which would have been achieved without the influence of the damage factors which are to be controlled. Since it is the aim of insecticide trials to achieve maximum control effectiveness, this assumption no longer holds true in the case of very high infestations by the *brown planthopper* or *tungro virus.* In this case, even several applications at the recommended rate cannot completely prevent the infestation.

Table 6 lists the *crop loss* in non-resistant varieties computed from trials conducted by IRRI and the BPI. The experiment stations concerned are located in different parts of the country with different agro-climatic characteristics. At IRRI, Bicol Rice Experiment Station (BRES) and Maligaya Rice Research and Training Center (MRRTC), rice is grown year-round so that pest reproduction potential is promoted, particularly in the case of brown planthopper and tungro virus. At the Visayas Rice Experiment Station (VRES), on the other hand, very little rice is grown during the dry season because of the uncertain irrigation situation. For this reason, occurrence of the tungro virus, in particular, is virtually ruled out. This explains the considerably lower average losses in VRES, which amount to only slightly over 20%, whilst at IRRI and BRES they amount to over 40%. The trial results at IRRI and BRES show yield differences each year, the minimum being 13.8%. The maximum *crop losses* of 91% occurred in MRRTC, but within some years no loss occurred at all. This also applies to VRES where the maximum *crop loss* amounted to 40.5% (Table 6). When the relative and absolute yield differences were computed, the plots treated with *Carbofuran* were dealt with separately (Table 6). This is necessary because Carbofuran, an insecticide of the carbamate group, has a yield-stimulating effect, producing yield increases of as much as 5–15%. Studies conducted by IRRI have shown that in the rainy season yield increases of as much as 1 ton can be achieved, provided the chemical is incorporated into the soil in granular form

Table 6. Crop loss due to insect pests and virus diseases in non-resistant varieties, based on results from pesticide evaluation trials, 1969–1981, in percent

Experiment station	Rainy season						Dry season					
	Without Carbofuran[a]			With Carbofuran[a]			Without Carbofuran[a]			With Carbofuran[a]		
	min	max	$\bar{x}$	min	max	$\bar{x}$	min	max	$\bar{x}$	min	max	$\bar{x}$
Laguna (IRRI)	35.9	60.5	44.2	21.5	84.1	49.5	13.8	55.7	43.8	27.6	60.6	44.8
Bicol (BRES)	26.5	74.7	43.6	29.6	85.9	60.8	34.8	64.3	44.5	21.7	39.9	28.8
Iloilo (VRES)	0	33.2	23.2	0	39.0	12.3	0	25.5	13.9	26.0	40.5	33.3
Nueva Ecija (MRRTC)	0	46.0	29.9	0	91.0	37.5	0	71.0	28.8	0	66.0	25.0
Average			35.2			40.0			32.7			33.0

[a] Crop loss in relation to the yields of the plots treated with Carbofuran

Source: Computed on the basis of the IRRI Annual Reports, the IRRI Insecticide Evaluation Reports (different issues) and the Annual Reports of the BPI Experiment Stations

prior to the planting of the rice. In the dry season, *crop losses* are lower than in the rainy season (Table 6).

The failure of Carbofuran to produce a growth-stimulating effect during the dry season is probably due in part to the higher use of fertilizer.

As regards the use of the results of pesticide evaluation trials for *crop loss assessments,* the following reservations must be made:

- At most of the experiment stations, rice is grown almost continuously. Pest reproduction potential is therefore not the same as under farm conditions.
- Pesticide evaluation trials are conducted on relatively small plots. Due to interplot interference, untreated plots, in particular, are consequently subject to an intensity of infestation unlikely to occur under farm conditions. *Crop losses* computed on this basis would therefore be overestimated.
- The variety IR-22 primarily used in insecticide trials does not correspond to the varieties used by the farmers. In this respect, also, *crop losses* would therefore be overestimated.

From the above it is clear that the results of pesticide evaluation trials cannot be used for computing current on-farm *crop losses.* They can, however, serve as indicators of the approximate order to magnitude of the *crop losses* likely to occur when the resistance of currently used varieties breaks down.

2.1.2 Trials Concerning Higher Levels of Crop Protection Intensity

The difference between the yields realized by the farmers and the yields achieved at experiment stations is referred to as Yield Gap 1, whilst the difference between the farmers' actual yields and the possible long-term average yields under field conditions is known as Yield Gap 2. The latter is attributable to an inefficient use of production inputs by farmers, both in technical respects and with regard to resource allocation. Yield Gap 2 therefore refers to the difference between maximum and optimum yield (Gomez 1977).

In 1974, IRRI started to set up trials for identifying and quantifying the factors determining this Yield Gap (De Datta et al. 1979). Starting with the customary farm input level and using the varieties actually used by farmers, the fertilizer, herbicide and insecticide levels in the test trials generally cover two further levels, i.e. an intermediate and high input level. In the case of crop protection trials (excluding weed control), the intermediate input level consists of one application of granules in the seedbed, two after transplanting and, whenever infestation occurred, one spray. For the high input level, granules were applied twice to the seedbed and as many as three times after transplanting. In addition, there is one prophylactic spraying in the transplanted field. As opposed to the method tested, on-farm insecticide input generally consists of one to two sprayings and occasionally, depending on the location, one application of granules. The yield difference between the fields worked by the farmer and the plot with the highest insecticide input level is then equal to the *residual crop loss* in the case of the customary on-farm crop protection measures. Table 7 lists the trial results obtained in four provinces during the period from 1975–1979. The yield

Table 7. Yield differences resulting from intensified crop protection measures in four Philippine provinces, 1975–1978

Province	Rainy season						Dry season					
	Absolute difference ($t\ ha^{-1}$)			Relative difference[a] (%)			Absolute difference ($t\ ha^{-1}$)			Relative difference[a] (%)		
	min	max	$\bar{x}$	min	max	$\bar{x}$	min	max	$\bar{x}$	min	max	$\bar{x}$
Laguna	0	1.37	0.688	0	27.8	15.47	0.7	1.0	0.795	11.75	19.5	15.28
Camarines Sur	0	0.545	0.269	0	13.15	6.25	0.2	0.9	0.45	4.88	17.0	9.16
Iloilo	0.3	1.6	0.725	6.88	27.12	13.76	0.1	0.6	0.3	2.1	10.72	5.9
Nueva Ecija	0.2	0.9	0.483	5.84	24.33	12.78	0.2	1.0	0.58	4.45	16.67	10.5
Average			0.541			12.06			0.531			10.21

[a] Computed from the difference between farmers' yields and yields after high-input level insecticide treatments
Source: Based on trial results from IRRI Research Paper Series No. 30 and IRRI Annual Reports 1978 (p. 298) and 1979 (p. 352)

increases range from 0 to 1.6 t ha^{-1}, corresponding to a relative maximum yield difference of 28%. On the average, with an approximately equal absolute difference of 0.5 t ha^{-1}, there is a relative difference of 12% in the rainy season and of 10% in the dry season. These differences are attributable to the different yield level and the lower infestation intensity during the dry season.

The relatively high losses in the Provinces of Laguna, Nueva Ecija and Iloilo are probably mainly due to the repeated occurrence of the brown planthopper. The lowest yield differences overall are found in the Province of Camarines Sur, a typical tungro virus infestation area, and are due to the fact that the majority of the farmers participating in the trials planted resistant varieties during the investigation period, so that virus infections did not occur. However, since there were no untreated control plots in the trials, the loss that would have occurred without any treatment *(potential loss)* cannot be ascertained. As a result, even these trials can hardly supply any useful data on crop losses. They do, on the other hand, convey an idea of the kind of yield increase which is possible to achieve by applying insecticides. However, since almost only *Carbofuran* granules were used, interpretation of the data is difficult, and the yield increases ascertained need not necessarily be the result of effective pest control.

By increasing the input levels of all three factors of production, the above trials can also be used to quantify the relative contribution of crop protection to the Yield Gap. If all the trials are taken into account, the contributions of fertilizer and crop protection in the rainy season are roughly equal. In the dry season, the productivity of fertilizer inputs is considerably higher than that of insecticide inputs. Weed control ranks last in both cultivation periods (Herdt 1979).

2.1.3 Trials Concerning Alternative Pest Control Methods

Further data for *crop loss assessment* can be derived from the trials conducted for developing specific pest control recommendations for the different regions (Litsinger et al. 1980). Analysis of these trials is concerned with:

- quantifying crop losses at the different plant growth stages,
- identifying the pests responsible for the crop loss,
- analyzing the present pest control methods used by farmers and evaluating the possibilities of introducing alternative pest control procedures,
- selecting a suitable control strategy,
- testing this strategy under field conditions in cooperation with selected test farms,
- assessing the economic feasibility of the technologies recommended.

Trials of this kind were started in 1976 and were initially limited to two provinces (Iloilo and Pangasinan). Subsequently such trials were also conducted in the Province of Nueva Ecija and became, in a modified form, part of the applied research activities of the pilot projects on integrated pest control in rice.

The *control measures* tested consisted of several sprayings at all growth stages, with one growth stage in turn always remaining untreated. In addition,

Table 8. Crop loss due to pests in resistant varieties, 1976–1980[a b] in t ha^{-1} and %

Planting period	Pangasinan						Iloilo						Nueva Ecija[c]					
	t ha^{-1}			%			t ha^{-1}			%			t ha^{-1}			%		
	min	max	$\bar{x}$	min	max	$\bar{x}$	min	max	$\bar{x}$	min	max	$\bar{x}$	min	max	$\bar{x}$	min	max	$\bar{x}$
Rainy season																		
– early planting	0	0	0	0	0	0	0	0.54	0.108	0	22.2	4.44	–	–	–	–	–	–
– late planting	0.47	1.36	0.87	11.22	24.16	19.89	0	0.69	0.413	0	24.2	12.23	0	1.7	0.57	0	29.02	9.67
Total, rainy season	0	1.36	0.58	0	24.16	13.26	0	0.69	0.223	0	24.2	7.36	0	1.7	0.57	0	29.02	9.67
Dry season	0	1.54	0.559	0	40.32	13.44	0	0	0	0	0	0	0	0.35	0.12	0	5.78	1.93
Total losses			0.557			13.32			0.2			6.54			0.34			5.8

[a] Only statistically significant yield differences are regarded as losses
[b] The mean values have been computed from all the individual values of the corresponding group
[c] Trials were not started until 1979
Source: IRRI Insecticide Evaluation Reports 1979 and 1980

alternative methods were tested, e.g. application of *Carbofuran* in granular form and use of control thresholds as a basis of the measures. One plot always remaining free of chemical treatments. In one test area, six to eight farms were generally used as replications of the different trial entries, or else several replications were set up on one and the same farm.

Crop losses can be computed for all the trials, i.e. for all input levels tested, the reference value always being the *potential loss* derived from the yield difference between the variant with the highest input level and the untreated plot. The corresponding analysis of results is summarized in Table 8. It will be noted that there are trials in all three provinces in which the yield differences between treated and untreated plots were not statistically significant. For these results, a 0 (zero) has been entered in Table 8. On the average, all planting dates or planting periods result in a crop loss which, depending on the area concerned, ranges from 200 to about 600 kg ha^{-1}, corresponding to 6–13% of the *potential yield*. In the Provinces of Neuva Ecija and Iloilo, losses are higher in the rainy period than in the dry period. In Pangasinan, on the other hand, they are lower in the rainy season. In the latter cultivation area, losses are highest, amounting to over 1.5 t, i.e. to 40% of potential crop yields. the highest absolute yield difference was ascertained in the Province of Nueva Ecija. However, due to the higher yield potential in this province, this difference corresponds only to a relative crop loss of not quite 30%. In the Province of Iloilo, losses are lowest, amounting to about 100 kg ha^{-1} at most. It is interesting to note that early planting during the rainy season results in lower losses than planting at a later date (Table 8).

If we compute an overall average from the mean values of the three cultivation areas, for all the trials listed in Table 8, we obtain a crop loss of 8.6%. On the basis of a yield level of 5 t ha^{-1}, this corresponds to 350 kg of rice.

To sum up, we can say that the trials for testing alternative pest *control measures* fulfill the preconditions necessary for valid *crop loss assessments*. However, two aspects pose a problem:

- In the case of trial variants with high insecticide input levels, yield-stimulating effects may be produced. Yield differences computed on this basis consequently do not correspond to the definition of *crop losses*.
- If the yield difference between treated and untreated plots is not statistically significant, the trial hypothesis must, strictly speaking, be rejected. In this case, which is not infrequent, the yield difference cannot be regarded as a loss without further analyses. The homogeneity tests (Puri and Mullen 1980) preceding the analysis of variance and a simultaneous comparison of the infestation and yield differences can then serve as further criteria of assessment.

2.1.4 Trials Conducted by the Author

The trials carried out in collaboration with the regional Crop Protection Centers follow the basic concept of the method outlined in Section 2.1.3. The inherent disadvantages of this method consequently also apply to the results of the trials described below.

Table 9. Crop loss due to pests, based on results from trials in farmers' fields, 1980–1981

Area	Iloilo				Nueva Ecija				Camarines Sur			
Cultivation period	Max yield $t\ ha^{-1}$	Yield of control plot	Difference[a] abs $t\ ha^{-1}$	Difference[a] rel %	Max yield $t\ ha^{-1}$	Yield of control plot	Difference[a] abs $t\ ha^{-1}$	Difference[a] rel %	Max yield $t\ ha^{-1}$	Yield of control plot	Difference[a] abs $t\ ha^{-1}$	Difference[a] rel %
1980 Rainy season	5.02	4.88	0	0	4.44	3.63	0	0	–	–	–	–
1980 Dry season	7.57	6.7	0.87	11.5	7.22	6.21	0	0	6.03	4.32	1.71	28.56
1981 Rainy season	6.76	7.01	0	0	3.97	2.88	1.09	27.46	2.65	2.46	0	0
1981 Dry season	4.68	4.17	0	0	–	–	–	–	4.69	3.98	0.79	18.64
Average				2.87				9.15				14.57

[a] When the maximum yield did not differ significantly from the untreated plot, a loss of 0 (zero) was entered

Source: Trials conducted in collaboration with the regional Crop Protection Centres

Expanding our original trial concept, we also included as variants the typical pest control practises used by the local farmers in the areas concerned. In addition, we measured infestation intensity at weekly intervals, whereas in the IRRI trials, assessments of infestation levels were made only on predetermined dates.

Trials were also carried out on the fields of the farmers using their own varieties. The results showed that in six out of a total of ten trials, no significant difference was found between the control measure aimed at providing maximum protection and the unsprayed field (Table 9).

Regarding the cases in which significant differences in crop loss did occur, the following must be noted: in the Nueva Ecija area, a considerable part of the *crop loss* in the 1981 rainy season was due to damage by rats just prior to the harvest, as a result of the rather late planting dates. In Camarines Sur, the resistance to the tungro virus of the IR42 variety used gradually broke down during the 1980 dry season and a *tungro virus* infestation of about 10% occurred on the untreated plots. The crop loss of 18% during the 1981 dry season is attributable to the high incidence of the rice leaf folder and the armyworm. Comparing this trial series with the previously mentioned IRRI trials, crop losses will be found to be almost identical, amounting to 8.9%.

2.1.5 Summary of Crop Loss Assessments in the Rice Crop of the Philippines

Assessments of crop losses due to pests in rice in the Philippines are based on very heterogeneous experimental material. This is partly due to the fact that the trials concerned were not primarily conducted for the purpose of *crop loss assessment* and can therefore serve merely as secondary sources for the purposes of the present study. Table 10 summarizes the data from the different sources.

Although, due to the heterogeneity of the data, a direct comparison of the individual *crop loss* is not possible, it will nevertheless be noted that there were some trials in which no losses occurred. This shows that infestation varied, both in time and from place to place. Furthermore, it is quite evident that the spread

Table 10. Crop loss due to pests in the Philippines summary of data from different sources

	Author/source				
	Pathak and Dyck	Pesticide evaluation trials	Yield constraints trials	Litsinger method	Own trials
Period	1968–72	1972–81	1975–79	1976–80	1980–81
Proportion of trials[a] without significant yield differences (%)	–	14.0	–	57.0	60.0
Max loss (%)	32.0	91.0	27.1	40.3	27.5
Average loss (%)	22.5	34.0	11.1	8.6	8.9

[a] Proportion of trials in which significant yield differences between treated and untreated were not found to exist

Table 11. Yield and infestation levels in test trials in Nueva Ecija, 1980 rainy season

	Yield (ha^{-1})	Leaf pests (%)	Leaf folder (%)	Stem borer (%)	Losses (%)
Maximum treatment	5.86	0.51	0.0	0.5	–
Untreated	4.166	2.816	6.0	1.0	2.02

Source: IRRI Insecticide Evaluation Report, 1980

of resistant varieties has reduced average *crop loss* and lowered the probability of significant yield differences occurring between treated and untreated plots.

Overall, the method of determining *crop loss* by means of a simple comparison of the yields of treated and untreated plots must be used with caution. Even if all the technical requirements of the trial are fulfilled, an interpretation solely of the yield data can often lead to wrong conclusions.

An example of this is presented in Table 11. Looking solely at the yields, we obtain a yield difference of 1.7 t ha^{-1}, corresponding to a crop loss of 29%. If we try to attribute this loss to a correspondingly high level of infestation, we find that the infestation intensity on the untreated plot is low. It is difficult to conceive that leaf damage of almost 3% due to leaf-feeding insects and 6% damage by the rice leaf folder can result in an overall loss of 29%.

Without further investigating the causes of such results, it follows that valid *loss assessments* can be arrived at only by taking into account the growth stages of the host plants and the infestation intensities detected in them.

It is only when these data are available that *crop loss* can be assessed by means of more sophisticated methods, i.e. by determining quantifiable infestation intensity/crop loss relationships, as already practised by Kirby and Archer (1926). With certain reservations, the data available for this study are suitable for *regression analysis* for computation of the infestation intensity/crop loss relationships or damage coefficients for the major pests, as we shall call them in the following.

2.2 Crop Loss Assessments by Means of Damage Coefficients

The assessment of Crop loss on the basis of damage coefficients presupposes that functional relationships exist between one or more infestation criteria and crop yield, i.e. loss in yield and quality. This mode of procedure is a rough simplification of a very complex biological system consisting essentially of three main components, namely the crop plant, the pest and the environmental influences. Each one of these components on its own in turn represents a subsystem, the structural elements of which determine the damage coefficient. As far as the pathogen is concerned, it affects the damage coefficient by such pest population characteristics as abundance (density), spread (distribution), morbidity (disease incidence), sex index (sex ratio), birth rate (fertility), constitution (capacity), age structure, etc. (Wetzel et al. 1980). As far as the crop plant is concerned, its particular growth stage is important both for the development of the individual yield components and with respect to its ability to compensate for the injury caused by the pest. By selecting an appropriate infestation criterium, the number of influencing variables can be reduced and the system thus simplified. Damage symptoms such as the percentage of dead hearts (stem borer) or the proportion of infested leaves (leaf pests) are the criteria currently used for infestation surveys in rice cultivation. The damage coefficient computed on the basis of these data represents so-called loss magnitudes which merely permit crop losses to be estimated, but are only of limited value for determining *control thresholds*.

In the following, the author will discuss the methodology of the available data and the results obtained which were tested for their utility within the scope of the present study.

2.2.1 Methodology

Before use is made of the regression analysis to compute damage coefficients, some basic questions must first be clarified. There are a number of possible ways of selecting both the exogenous and the endogenous variables. Theoretical considerations are also necessary to determine the type of function to be used, i.e. deciding whether linear or non-linear relationships exist between the exogenous and endogenous variables. Finally, selection of the data also has to be geared to the purposes for which the crop loss relationships are to be determined. To this end, specific processing of the data partially collected for other purposes is often necessary. Estimates of infestation intensity, for instance, which are expressed in terms of scales of assessment, have to be converted into percentages. Similarly, crop yields given in absolute units have to be converted into crop loss percentages.

2.2.1.1 Selection of Exogenous Variables

If we classify the *regression models* according to the exogenous variables selected, we obtain two types of model:

a) Models based on the hypothesis of the existence of one or several growth periods which are critical for yield losses caused by any particular pest.
b) Models in which the area under the infestation curve represents the dependent variable (Van der Plank 1963).

The advantage of model (b) as compared with model (a) is that the former takes more account of the time factor, thus allowing for the fact that for example identical infestation maxima can be due to very different disease or pest progress curves and may consequently also result in different losses. In the case of model (a), an attempt is made to take into account the development in time by including as exogenous variables in the regression model the infestation intensities at different points in time. However, this often results in the problem of *multicolinearity* of the exogenous variables so that the parameters of these variables have extremely high standard errors and are not significant, although their overall influence is significant (Schneeweiß 1978). Verification of the results of the model using independent data is generally impossible since the regression parameters determined can assume implausible values. When there is a high degree of correlation between the exogenous variables, it is generally necessary to dispense with one of the two variables and to run the model with the infestation level most highly correlated to the loss.

A further possibility of modifying model (a) is to use the cumulated infestation intensities as exogenous variables. Although this rules out the problem of multicollinearity, it does mean that several infestation surveys conducted at regular intervals are always required.

In view of the nature of our data, critical point models for one or more pests will be used in this study and, whenever possible from a statistical point of view and necessary in biological respects, several infestation stages will be included (multiple point models).

2.2.1.2 Selection of Endogenous Variables

As regards selecting the independent variables in *crop loss models,* there are also two possibilities. Firstly, infestation intensity can be related to the measured yield per unit area and, secondly, *crop loss* expressed in terms of the percentage difference from the potential yield can be used as an explanatory variable in the *regression model.*

In the case of data based on exact single-factorial trials, the first possibility does not pose any great problems, since the yield differences are attributable only to the differences in infestation intensity. Thus, further processing of the data is not necessary. However, this method of computing damage coefficients presupposes that the data used have either been obtained from a field trial or that further explanatory variables are incorporated. In the case of *loss profiles,* the latter is definitely necessary to explain the differences in yield between different farms or regions (Pinnstrup-Anderson et al. 1976). The value of damage coefficients derived from the analysis of single-factorial trials is limited since they always apply to only one production period. Therefore their use both for *loss assessment* and as a major component of *control thresholds* should, strictly speaking, be restricted to a posteriori considerations.

On the other hand, if *crop loss* is selected as a dependent variable, data from several cultivation periods can be employed. To this end it is necessary to fix a basic yield for all trials. This is generally the yield of non-infested or at any rate almost non-infested plots.

2.2.1.3 Determination of the Potential Yield

The basic yield, also referred to as *potential yield,* is the yield which would have been achieved without the influence of the pests. There are essentially two ways of determining the *potential yield:*

a) By extending the regression line up to where it cuts the ordinate. The point of intersection then corresponds to the yield under conditions of zero infestation, i.e. to the *potential yield.* This method presupposes that the correct function has been found and that any possibly existing non-linear relationship can be taken into account by suitably transforming the variables. Whether this method is statistically justified can be tested by means of the test value for the *Y* intercept β_0. In addition, the confidence interval (CI for β) can be computed, namely $CI = \beta_0 + st$ (Sachs 1971)

where s = standard error of β_0,
t = value from the t-distribution for $(n-1)$ degress of freedom.

The lower limit of the β_0-CI interval for a significant β_0 can then be used as a basis for computing the percentage loss. This appears to be justified in as much as only values below this limit represent a loss from a statistical point of view. If β_0 were to be used as the basis, then any yields lower than β_0, but still within the confidence interval of β_0, would wrongly be regarded as losses.

b) In the method more frequently used in practice, treatments aimed at minimizing infestation intensity are included as trial entries in field trials so that these yields can be regarded as the *potential yield.* However, the use of chemicals which possess a yield-stimulating effect (e.g. Carbofuran) may lead to misinterpretations.

The use of copper-, zinc- or sulphur-containing fungicides may obviate a possibly existing nutrient deficiency and thus produce a positive effect on yields (Kelber 1975). The date listed in Table 11, in which a negligible difference in infestation levels has produced a yield difference of about 30%, seem to indicate that some of the insecticides used for spraying may have such an effect. On the other hand, in the case of extremely high infestation intensities, there may be not a single plot on which pest incidence can be regarded as still being below the *physiological damage threshold,* so that the loss computed would be an underestimation. Since in the case of single or multi-pest infestations it is generally impossible, even with intensive treatment, to obtain entirely infestation-free plots, this method always requires an a priori assumption, based on pragmatic values or prior trials, regarding the *physiological damage threshold.*

2.2.2 Selection of the Correct Type of Function

In the majority of studies dealing with this type of problem, a linear relationship is assumed to exist between infestation intensity and crop loss. As far as the field of entomology is concerned, the use of linear functions goes back to a model first described by Tammes (1974) (Fig. 2). This model is based on a sigmoid curve applicable to many biological relationships. Mathematically, this curve can be expressed as a lagged logistic function of the following type (Förster and Rönz 1979):

$$Y = \frac{a}{\frac{b^c}{x}} \quad \text{or} \quad \log\frac{a}{y} - 1 = c\log b - c\log x.$$

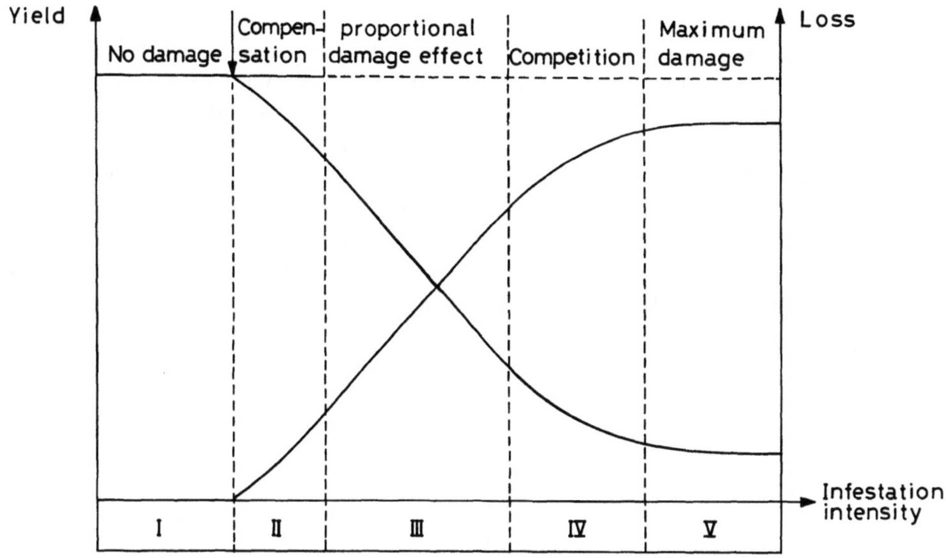

Fig. 2. Relationship between pest infestation and yield loss

From Fig. 2 it can be seen that the graph can be subdivided into several sections. The trial results of the individual years are reflected by one or more of these sections. Part 1 of the function describes the range over which pest infestation does not produce a yield-reducing effect, i.e. the range of infestation levels up to the *physiological damage threshold.* The physiological damage threshold denotes the level at which any further increase in infestation will result in a yield decrease. In other terms, this is the point at which the production function departs from its shape without pest influence.

Section 2 of the function marks the beginning of yield decline. Due to the weakening resistance of the plants, yield losses increase. Frequently – as can be shown by the example of stem borer damage (Ishijura 1969) – this can be represented by squaring the independent variables. With the further increase of infestation intensity, the power of compensation of the plants becomes completely exhausted, so that infestation level and yield are now inversely proportional. The part of the curve where a linear relationship exists makes up the major part of the infestation range and can consequently be most easily verified by means of empirical data. The range of high infestation intensities can be represented by means of an exponential curve. The shape of this part of the curve is due to competition effects within the pest population which occur when population densities are high. Once the peak has been reached, the population breaks down and no further yield decreases occur. This fifth part of the function, like the first part, assumes the form of a line parallel to the abscissa.

Plotting this function (Fig. 2) with crop loss in percent as a dependent variable facilitates the use of linear functions. The range below the *physiological damage threshold* is represented by the negative section of the axis. The effect of plant compensation is taken into account by transformation of the variables when the basic yield for the respective initial function is determined. The influence of plant compensation is thus expressed by the value of the Y intercept. From an economic point of view, the error which occurs when extrapolating the line to the range of high infestation levels can be neglected, since in the case of extremely high losses, further changes in the optimum control strategy are not to be expected. However, it is also possible to take account of the actual shape of the curve by plotting yield as a logarithm of infestation intensity.

The theoretical conclusions arrived at by Tammes (1974), were also verified by means of an empirical study. A model was developed for determining the crop losses caused by different plant diseases. Mathematically, this model can be described as follows (Madden et al. 1981):

$$L = 1 - \exp \{- [(x - d) / b]^s\},$$

where: L = crop loss in %,
x = measure of infestation intensity,
d = physiological damage threshold (disease threshold),
b = slope parameter,
s = parameter describing the form of the function.

Madden et al. (1981) showed that for a number of different plant diseases, this model, when applied to independent data, results in a more exact presentation of the quantitative relationships than other, simpler types of function. The data available for the present study do not meet these requirements since depending on the type of infestation situation concerned, only parts of the model curve can be plotted. Furthermore, in the studies carried out to date in this field, mainly linear functions have been employed (Kelber 1975). For rice diseases, too, *(rice blast* or *sheath blight)* simple, linear functions are generally used. In the present study, linear functions have also been employed for most of the pests and, whenever necessary for statistical or biological reasons, non-linear tranformations have been carried out. When selecting the type of function to be used, it is therefore not sufficient to consider statistical criteria alone. It is also necessary to check the plausibility of the function, as shown by the following example: In 1972, the first comprehensive trial was carried out to determine the losses caused by certain *stem borer* species [*C. suppressalis* (Wlk.) and *T. incertulas* (Wlk.)] by means of yield and infestation measurements (Gomez and Bernardo 1974). In this trial, yields and infestation levels were measured for the individual hills which in turn were classified according to varieties and fields. For data analysis, the individual hills were then grouped into infestation classes of 2% each and an exponential curve $Y = ax^{-b}$ was fitted to present the functional relationship between yield per hill and the percentage of infestation *(white heads)* measured 90 days after transplanting (Fig. 3.).

Trying to explain from a biological point of view the function shown in Fig. 3, we will be led to suspect that the exponential shape of the curve is caused only by the extreme yield values in the lower infestation range. The shape of the curve corresponding to these values does not fit the function $Y = ax^{-b}$, nor does it seem logical from a plant physiological point of view.

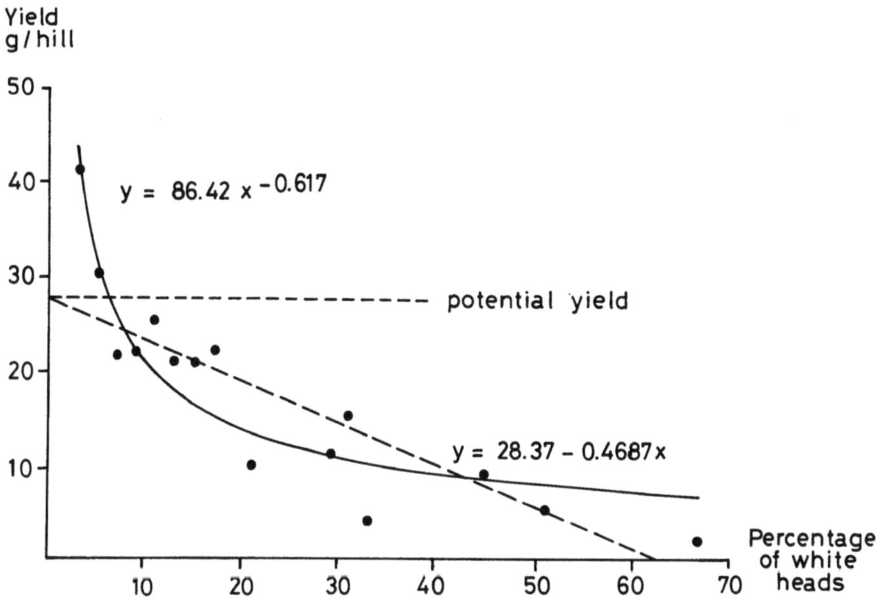

Fig. 3. Relationship between hill yield and percentage of white heads

If we assume that there is no causal connection between peak yields and low infestation intensity, we can also compute a linear regression between both variables.

$$Y = 28.37 - 0.4687\,X^{**} R^2 = 0.716,$$

where Y = crop yield in g per hill, and
X = white heads in %.

With a degree of determination of 0.716, the function thus computed is only marginally worse than the exponential function given by the authors. The regression parameters of the linear function are nevertheless significant. From the confidence interval of the Y intercept, we can then compute the basic yield required for determining the percentage loss, the value arrived at being 27 g per hill. Infestation intensity can therefore be regressed on the dependent variable "percent loss":

$$Y = 1.839 + 1.5949\,X^{**} R^2 = 0.84,$$

where Y = crop loss in %,
X = white panicles.

The coefficient of determination of this function is clearly higher because the conversion into loss percent results in correction of the data with respect to the extreme values in the lower range of infestation intensities.

The above comments are intended to show that even if the shape of the curve is optimum from a statistical point of view, this does not necessarily mean that it reflects the actual relationship between infestation intensity and yield or crop

Table 12. Overview of available damage coefficients for diseases and insect pests of rice in Asia

Pest/disease	Type of function	Variables		Country	Author
		Endogenous	Exogenous		
Pyricularia oryzae	$Y = 0.7895 + 0.4474X$	Crop loss in %	Percentage of infested tillers at harvest time	India	Mathur et al. 1964
Sheath blight	$Y = 6629.24 - 151.43X$	Yield in kg ha^{-1}	1–9 Scale of assessment	Philip-pines	IRRI Ann. Report 1975
Xanthomonas oryzae	$Y = 92.69 - 0.445X$	Yield in %	Infestation intensity in %	India	Reddy et al. 1979
Brown planthopper	$Y = 10.898 + 0.126X_1 + 0.47X_2 + 0.3 + 6X_3$	Crop loss in %	X_1 = % of dead panicles X_2 = degree of panicle damage X_3 = degree of lodging of the infested plants	Japan	Nomura 1951
Brown planthopper	$Y = 10 + 0.03X$	Crop loss in %	X = No. of insects per net sweep	Japan	Kisimoto 1975
Brown planthopper	$Y = 5464.96 - 1126.17X$	Yield in kg ha^{-1}	X = Damage index 1–5	Japan	Sogawa and Cheng 1979
Stem borer (different rice varieties)	$Y = 44.1X^{-0.354}$ $Y = 86.42X^{-0.617}$ $Y = 1.47X^{1.553}$	Yield in kg ha^{-1} Yield in kg ha^{-1} Crop loss in %	X = % of white heads X = % of white heads X = % of white heads	Philip-pines	Gomez and Bernardo 1974
Stem borer	$Y = -0.6 + 11.6 \log X$	Crop loss in %	X = % of white heads and onion shoots	Bangla-desh	Catling et al. 1978
Rice bug	$\log Y = 1.8553 - 0.4717 \log X$	Crop loss in %	X = No. of panicles per rice bug	Philip-pines	IRRI Ann. Report 1975

loss. Whenever the data available are rather limited, it seems advisable not to rely on statistical considerations alone, but also to consider plausibility when deciding on the mode of determining the basic yield. Otherwise a curve may be obtained the shape of which is impossible to explain in biological respects. This may lead to erroneous results in economic computations, notably in the lower to medium infestation intensities.

Apart from the trials conducted by Gomez and Bernardo (1974), not many attempts have been made in Philippine rice cultivation to develop models for assessing losses on the basis of infestation intensity. In Table 12, a listing is given for the Asian region of the models known to the author. For further details in this respect, refer to the literature indicated in the table. For the purposes of the present study, it will suffice to say that due to the kind of exogenous variables selected, the models listed are unsuitable for *loss assessments* or computation of *control thresholds* for rice diseases and they are suitable only to a very limited degree for insect pests. Furthermore, the majority of the models are based on only a few observations and generally on only one cropping period. As regards type of function, linear relationships predominate, thus confirming the model assumptions, regarding the most frequent range of infestation intensities.

2.2.3 Presentation of Available Data

As a data basis for the determination of *damage coefficients* secondary sources, as well as data from trials conducted by the author specifically for this purpose were used. The secondary data available to us were from greenhouse and pot trials, from the official pesticide evaluation trials and from trials for determination of alternative pest control techniques. These sources are partly mentioned in Chapter 3.2.

To supplement the above data, the following field trials were conducted in collaboration with the regional Crop Protection Centres: (1) trials in which the control strategies commonly used by the farmers were used as trial variants, and (2) trials with a number of marked, untreated individual plants (hills) per field.

As a further possibility, data from the observation fields set up on farms could have been used. However, since only in a very few cases were marked differences in infestation intensity actually ascertained between treated and untreated fields, this source of data was not used.

In all the data available, differentiation must always be made between those for resistant and non-resistant varieties. While the greenhouse and plot trials were partially conducted with non-resistant varieties, those varieties were used alone in the pesticide evaluation trials. The varieties used in the remaining trials were resistant to the *tungro virus* and the *brown planthopper.*

2.2.3.1 Data for Non-Resistant Varieties

From 1974 onwards, further trials for determining crop loss due to rice pests were conducted at IRRI. The results of these trials were published in various studies. For our purposes, the paper by Dyck et al. (1977) is of interest.

Table 13. Data for the determination of damage coefficients for non-resistant varieties[a]

Pest/disease	Trial location	Trial years	No. of trials	No. of observations (*n*)	Variety
Stem borer	Iloilo	1970, 1971	3	21	IR-8, IR-5
Stem borer	Iloilo	1975–1978	4	48	IR-22
Tungro	Bicol	1972–1980	12	125	IR-22
Tungro and brown planthopper	Bicol	1977–1980	5	56	IR-22
Brown planthopper	Bicol	1978	3	26	IR-22
Tungro	Nueva Ecija	1973–1977, 1971	6	47	IR-22
Tungro and brown planthopper	Nueva Ecija	1980–1981	3	36	IR-22
Brown planthopper	Nueva Ecija	1977, 1980	4	50	IR-22
Stem borer	Nueva Ecija	1967–1970	5	51	IR-8 etc.

[a] The data not found in the reports listed below was supplied by the staff in charge of the trials concerned: Mr. F. Olivares (MRRTC), Mr. C. Dancel (BRES) and Mrs. D. delos Santos (VRES)

Source: Annual reports of the experiment stations VRES (Iloilo), BRES (Bicol) and MRRTC (Nueva Ecija)

Although the findings arrived at are based on only a few different intensity levels of artificially produced infestation or loss respectively, the data are nevertheless suitable for regression analyses, since the infestation levels generated cover a wide range of infestation intensities.

Extensive data are collected within the scope of the official pesticide evaluation trials, but this material is not always suitable for loss assessment. In collaboration with the chemical industry, various new pesticides or pesticide combinations (mainly insecticides) are tested together with products already recommended. As a result, the range of infestation levels obtained depends to a large extent on the particular pesticides selected. The plots are laid out according to the randomized block design and include one untreated control plot.

The insecticides selected are applied according to a rigid schedule, i.e. a total of four times at 15-day intervals, with the first application taking place 10 days after transplanting. Infestation surveys covering the main pests are carried out during the critical development phases. During the vegetative phase, infestation by *rice whorl maggot, stem borer, green leafhopper* and *brown planthopper* is measured, and towards the end of this growth stage, the first measurement is made of *tungro* infestation. With the exception of the *rice whorl maggot,* the same assessments are also made during the generative growth phase. Before plot yields are determined, the damage caused by the *stem borer* is assessed once more *(white heads).*

Table 13 lists the trials carried out at three experiment stations of BPI and IRRI. The trials have been broken down according to major damage factors and indicate the main areas and years of pest occurrence.

The most extensive material is available for the tungro virus. Almost 40% of the trials deal with this disease, most of these trials being conducted at BRES. The *brown planthopper* accounts for somewhat more than 20% of the trials and, in roughly the same number of trials again, virus and planthopper occurred jointly. Up to the middle of the 1970's, the *stem borer* was the main damage factor and, in particular at VRES, it is still the dominant pest today.

By computing the damage coefficients for non-resistant varieties, it is possible to estimate the losses likely to occur if the resistance of the current varieties breaks down, the assumption being that *damage coefficients* would be identical.

2.2.3.2 *Data for Resistant Varieties*

In the preceding chapters, the trial series conducted according to the Litsinger et al. (1980) method has already been referred to several times. To a limited extent, data from these trials can be used for regression analyses. However, in the

Table 14. Data for the determination of damage coefficients for resistant varieties

Pest	Type of trial	Trial location	Trial years	No. of trials	No.[a] of samples (n)	Variety
Rice leaf folder	Litsinger method	Iloilo Pangasinan Nueva Ecija Laguna	1977 1979 1980	5	30	IR-36
Rice caseworm	Litsinger method + author's own trials	Iloilo Pangasinan Nueva Ecija	1978 to 1981	4	31	IR-36
Army worm/ cutworm (veg. phase)	Author's own plot trials	Nueva Ecija	1980 to 1981	3	22	IR-36 IR-42
Army worm/ cutworm incl. rice leaf folder	Author's own plot trials	Bicol	1980 to 1981	2	16	IR-42 IR-54
Rice whorl maggot	Individual plant trials	Iloilo	1981	1	300	IR-36
Rice stem borer and others	Individual plant trials	Iloilo	1981	1	300	IR-36
Rice bug	Litsinger method	Iloilo	1977, 1978	2	12	IR-36

[a] No. of trial plots

Source: Trials in collaboration with the regional Crop Protection Centres and IRRI Insecticide Reports 1979 and 1980

trials concerned, infestation assessments were only made at the time of the expected population peaks, which generally occur during the critical development phase, although it is impossible to check this hypothesis. In the trials carried out by the author in collaboration with the regional Crop Protection Centres, this fact was taken into account and assessments were consequently made at weekly intervals. The data from these trials are given in Table 14.

In view of the slight infestation levels in the plot trials in the Iloilo investigation area, a trial with individual plants was additionally set up. On a 600 m^2 field, to which no chemical treatments were applied at all during the entire plant growth period, 300 marked individual plants at the centre of the field were assessed at weekly intervals. The assessment included counting the number of tillers per hill.

The influence of weeds was eliminated through timely hand weeding. In addition to infestation intensity, the grain yield of the individual plants was measured. In this way it was possible, in contrast to the plot trials, to meet the statistical requirements (even in the case of only a slight overall infestation intensity) by measuring the level of attack on individual, relatively heavily infested plants. This trial was carried out once in the rainy season and once in the dry season. However, the overall number of data obtained was nevertheless limited and it was consequently impossible for separate computations of damage coefficients to be carried out for the different areas. In the computations of crop loss relationships, all the trials in which a similar infestation structure had occurred were therefore grouped together (Table 14).

2.2.4 Sorting of Available Data

Due to the short trial period of only 2 years, the trials set up to determine crop loss relationships did not provide sufficient statistically significant data for all the major pests. Furthermore, the low infestation levels prevailing during the trial period were disadvantageous as far as our objective was concerned. For this reason, use was made of the secondary data already referred to above. But even this data did not suffice for constructing functions – differentiated according to investigation areas – for all the pests. Therefore it was decided to group together all the trials in which the same pest had occurred as the main damage factor.

The method pursued was to compute the *potential yield* for the individual trials as described under Section 2.2.1.3. From these values, the percentage loss was then determined and employed as dependent variable in the function constructed on the basis of data from several trials.

Whenever it was obvious that growth-stimulating effects had been produced, the set of data from the plots treated with Carbofuran were excluded from the analysis. Growth effects could be assumed to have occurred whenever infestation levels were almost equal, but yields were found to differ significantly. Similarly, trials in which no significant differences could be detected between treated and untreated plots were not included in the analysis. This mode of procedure was modified only in the case of data from plot trials conducted by the author at the same site throughout.

In the case of many trial entries, the values obtained were highly concentrated in the lower to medium range of infestation intensities due to the fact that many similar and therefore not significantly different variants had been used. For this reason only the pairs of data from the completely treated and from the untreated plots were used for the regression analysis. In the case of the greenhouse and cage trials, the data were used unchanged as obtained from the samples, and a regression equation was then computed for each trial.

In the individual plant trial at Iloilo, all the hills were assessed according to phenotypical criteria during the three main growth stages (vegetative, generative and ripening phase). If, on the basis of this assessment, any hill had to be regarded as greatly differing from the norm, it was excluded from the analysis. The mean yield of non-infested and only slightly infested plants was used as the basic yield.

For the *rice whorl maggot,* the infestation data from the secondary sources were generally expressed in the form of scales of assessment so that assessment and measuring problems had to be taken into account in the case of the endogenous variables. It was therefore necessary to convert the scale values into percentages. This was done by taking the mean value of each interval as infestation percentage. For leaf pests, only infestation frequency – measured in terms of the number of infested leaves – was used in some of the trials. There were also a number of cases where the infestation intensity, although expressed as a percentage, was in actual fact an estimated figure. Both types of figures were used equally as percent infested leaves for the computation of the respective functions, since in practice both are considered admissable and a correlation can therefore be assumed to exist between them.

2.2.5 Determination of Damage Coefficients by Means of Simple and Multiple Regression Analysis

The existing data permitted the construction of regression equations of crop loss and infestation intensity for resistant and non-resistant varieties. The results will be grouped according to type of data used. First of all, the results of the greenhouse and cage trials will be presented. These trials will be evaluated particularly in view of determining the physiological damage threshold. They are not very suitable for the computation of *damage coefficients*. Next, the functions for stem borer, tungro and brown planthopper, computed from the data of the insecticide trials, will be presented. In the case of the brown planthopper, only the infestation during the generative growth phase could be taken into account.

As regards functions for resistant varieties, these will be constructed for individual pests using the data from the field trials. The infestation intensities during the particular development phases regarded as critical will be employed as independent variables. The data from the individual plant trials, on the other hand, can be used to construct multiple regression equations for several jointly occurring pests.

2.2.5.1 Greenhouse and Cage Trials

Functions constructed on the basis of data from greenhouse and cage trials should be regarded only as rough indications of crop loss relationships. Strictly speaking, the method used to produce the damage represents a consideration based on one selected time since under natural conditions damage is the result of a development in time. Population dynamic aspects are consequently not taken into account and the coefficients determined are unsuitable for the computation of control thresholds. At most, the results can provide information on the degree of pest tolerance of the rice plant.

The results of the regression analysis are presented in Table 16. Functions (3)–(8) permit a purely mathematical determination of the *physiological damage threshold.* Mathematically, the physiological damage threshold corresponds to the value of X (X = infestation) for $Y = 0$ (Y = crop loss). For example, if we set $Y=0$ in function (3), then

$$X = \frac{4.566}{0.7752} = 5.89.$$

The result shows that a stem borer infestation of an intensity of up to roughly 6% "dead hearts", setting in 40 days after seeding, would not produce any loss. However, in the case of function (3), the result cannot be regarded as statistically significant since the t-value computed for the Y intercept is not statistically significant. For this reason it seems necessary to indicate the confidence intervals for the computed threshold values.

Table 15 therefore gives the physiological damage thresholds and their confidence intervals for the functions listed in Table 16. It will be seen that the

Table 15. Physiological damage thresholds for stem borer, rice leaf folder and rice bug

Pest	Sampling unit	Mean value[b]	Confidence interval	
			Lower limit	Upper limit
Stem borer 40 DAT[a]	Percent dead hearts	5.89	0	11.8
Stem borer 60 DAT	Percent dead hearts	2.60	0	5.2
Stem borer 80 DAT	Percent dead hearts	2.00	0	4.4
Rice leaf folder 77 DAT	Percent infested leaves	7.7	6.5	8.8
Rice leaf folder 84 DAT	Percent infested leaves	7.3	4.5	10.1
Rice bug	No. m^{-2}	0.36	0	0.8

[a] DAT = days after transplanting
[b] On the basis of the respective sampling unit; calculations based on Table 16

Table 16. Functions for estimating crop loss due to pests, data obtained from greenhouse, individual plant and cage trials

No. of function	Pest	Type of function regression parameters	Exogenous variables	R^2 F value	Type of trial
1	Stem borer	$Y = 0.188 + 0.4075X$	X = Percent dead hearts 40 DAT[a]	0.96*** 81.04	Greenhouse
2	Stem borer	$Y = -1.095 + 2.0703X$	X = Percent dead hearts 60 DAT	0.994*** 531.02	Greenhouse
3	Stem borer	$Y = -4.566 + 0.7752X$	X = Percent dead hearts 40 DAT	0.843** 21.58	Individual plants in the field
4	Stem borer	$Y = -4.216 + 1.6338X$	X = Percent dead hearts 60 DAT	0.896** 34.75	Individual plants in the field
5	Stem borer	$Y = -3.701 + 1.8014X$	X = Percent dead hearts 80 DAT	0.931** 54.22	Individual plants in the field
6	Rice leaf folder	$Y = -14.00 + 1.82X_1$	X = Percent infested leaves 77 DAT	0.994*** 356.1	Greenhouse
7	Rice leaf folder	$Y = -14.47 + 1.986X_1$	X = Percent infested leaves 84 DAT	0.898** 35.34	Greenhouse
8	Rice bug	$Y = -0.92 + 2.523X$	X = No. of rice bugs per m^2	0.974*** 116.84	Cage trials

[a] DAT = days after transplanting; the endogenous variable is always Y = percent crop loss
Source: Computed on the basis of Dyck et al. (1977)
***: Significant at $P \leq 0.001$
**: Significant at $P \leq 0.01$

physiological damage threshold for the stem borer decreases as the plant growth period proceeds, because if the damage is produced 80 days after seeding, a yield decrease already occurs when infestation intensity exceeds 2%. However, attention must be drawn to the large confidence intervals which lend a high degree of uncertainty to this physiological damage threshold.

The values for the rice leaf folder possess a higher statistical significance. In this case, an infestation intensity of up to 8% can be tolerated. As regards the rice bug, the data show that crop loss occurs when the infestation intensity exceeds 36 bugs per 100 m^2.

2.2.5.2 Insecticide Trials

The data from the insecticide trials permitted statistically significant functions to be derived for all important damage factors in non-resistant varieties, e.g. IR-5,

Table 17. Functions for estimating crop loss caused by stem borer, brown planthopper and tungro, non-resistant varieties

Pest (no.)	Location (variety)	Type of function regression parameter	Exogenous variables	Confidence intervals		R^2
				Intercept	Regression coefficient	ε value
1. Stem borer	VRES (IR-5) (IR-8)	$Y = -3.89 + 1.4003X_1 + 1.6167X_2$	x_1 = Deadhearts in percent at reproductive growth phase	$-6.608 \leq a_0 \leq -1.772$	$0.464 \leq b_1 \leq 2.336$ $0.871 \leq b_2 \leq 2.362$	0.68 13.76
2. Stem borer	VRES (IR-22)	$Y = -5.148 + 0.414X$	x = Whiteheads in percent, taken at 10 days before harvest	$-7.953 \leq a_0 \leq -2.361$	$1.7558 \leq b \leq 2.327$	0.82 207.27
3. Tungro and brown planthopper	BRES (IR-22)	$Y = -1.29 + 0.1144X_1 + 0.7415X_2$	x_1 = No. of hoppers per hill at 60 days after transplanting x_2 = Percent of hills attached by growth stage during reproductive phase	$-0.379 \leq a_0 \leq 2.959$	$0.0628 \leq b_1 \leq 0.1659$ $0.105 \leq b_2 \leq 0.847$	0.89 238.87
4. Tungro	BRES (IR-22)	$Y = 4.58 + 0.783$	x = Percent of hills infested by tungro, reproductive phase	$1.636 \leq a_0 \leq 7.524$	$0.688 \leq b_1 \leq 0.878$	0.75 305.1
5. Brown planthopper	RRES (IR-22)	$Y = 4.024 + 0.00556X$	x = No. of hoppers per hill, 60 days after transplanting	$-1.722 \leq a_0 \leq 9.77$	$0.03 \leq b \leq 0.812$	0.46 20.04
6. Stem borer	MRRTC (IR-22)	$Y = 6.229 + 2.335X_1 + 1.7396X_2$	x_1 = Percent deadhearts, reproductive phase x_2 = Percent whiteheads, 10 days before harvest	$-9.736 \leq a_0 \leq -0.7224$	$1.6499 \leq b_1 \leq 2.8172$ $1.0188 \leq b_2 \leq 2.4604$	0.64 42.68

7. Tungro	MRRTC (IR-22)	$Y = -113.619 + 104.399 \log (x+10)$	x = Percent of hills infested by tungro	$-127.007 \leq a_0 \leq 100.161$	$95.2379 \leq b_1 \leq 113.4403$	0.92 533.74
8. Tungro and brown planthopper	MRRTC (IR-22)	$Y = 6.806 + 0.1075 X_1 + 0.0387 X_1 X_2$	x_1 = No. of hoppers per hill, 60 days after transplanting x_2 = Percent of hills infested by tungro	$4.0039 \leq a_0 \leq 9.6081$	$0.475 \leq b_1 \leq 0.1975$ $0.0294 \leq b_2 \leq 0.048$	0.77 51.4
9. Brown planthopper	MRRTC (IR-22)	$Y = 2.133 + 0.1657$	x = No. of hoppers per hill, 60 days after transplanting	$0.3618 \leq a_0 \leq 4.6278$	$0.0461 \leq b_1 \leq 0.2138$	0.44 36.75

IR-8 and IR-22 in particular. Although a satisfactory coefficient of determination was not obtained for all functions, the regression coefficients were found to be statistically significant in every case, as shown by their test values (Table 17). All the functions computed for the stem borer have regression coefficients of a magnitude >1, i.e. an infestation intensity of 1% results in a loss of over 1%. This is attributable to the fact that the number of "dead hearts" counted during the generative and ripening periods does not exactly reflect overall infestation development at the particular time. Under tropical rice farming conditions, this applies all the more, since pest population generations can be expected to overlap. When the first infestation assessment of the vegetative phase was included as a further variable, it resulted in a lower precision of estimation of the functions in every case. An early stem borer infestation consequently does not permit conclusions to be drawn regarding the resulting crop loss (Ishijura 1969). During this growth phase, the plant is still capable of producing additional tillers and, furthermore, the infestation does not occur only on the tillers subsequently developing into panicles, but also on unproductive tillers. In the case of all three functions, there is a statistically significant *Y* intercept with a negative sign. This confirms the assumption that a *physiological damage threshold* exists, even in the case of an infestation occurring during the generative growth stage. These threshold values would consequently correspond to roughly 2.8% dead hearts in the case of the function for the varieties IR-5 and IR-8 (Iloilo) and to roughly 2.3% in the case of the function for the variety IR-22. Since these functions are based on trial data covering several years, it can be assumed that a stem borer infestation of 2–3% dead hearts during the generative growth phase will not produce a measurable effect on crop yields. Although the estimated functions for crop loss due to tungro at BRES and MRRTC differ in shape, they lead to similar results in the range of high infestation intensities. In the case of BRES, the positive *Y* intercept of the function indicates a progressive increase in crop loss in the range of low infestation intensities. Thus, a tungro infestation of 10% would correspond to a crop loss of 12.4% while a 30% infestation would only result in a crop loss of 28%. The function based on data from MRRTC is proportional to the logarithm of infestation intensity. This means that a tungro infestation of 10% produces a crop loss of 22% and a 30% infestation, a crop loss of 53%. Although the fact that the infestation measurements were carried out by different persons may affect the result of the regression analysis, the differences can also be due to epidemiologic reasons. Overall, the functions for the brown planthopper have a low statistical quality. The fact that infestation development differs greatly in the different years reduces the value of punctual infestation surveys. The number of planthoppers per hill at a certain time during a particular growth stage is not per se a satisfactory measure of pest population.

The regression coefficients of both functions were nevertheless statistically significant, although clearly different. In the case of both functions, the *Y*-intercepts have positive signs, but are not significantly different from zero. If we neglect the *Y*-intercept of a function (Heady and Dillon 1964), then at BRES, 20 BPH per hill would cause a loss of only slightly more than 1%. At MRRTC, the same infestation intensity would result in a loss of 3.3%.

Trials in which both the brown planthopper as well as the tungro virus occurred were grouped together for the construction of functions. This seems plausible since an overlapping of infestation periods and damage effects can be expected to produce antagonistic or synergistic effects on both damage factors. In the regression analysis, this interaction is expressed in a multiplicative term of both exogenous variables.

However, when the corresponding function was computed for BRES, it was not possible to verify this assumption, since inclusion of a multiplicative term reduced the precision of estimation and reduced the test value of the variable X_1 to below its tabular value for $P \leq 0.05$. This result indicates that tungro was the dominating damage factor in the development of the crop loss. Overall, the function with the two pests as exogenous variables is statistically significant and has a coefficient of determination of 0.891. The Y intercept of the function is not statistically significant and is therefore ruled out for the purpose of loss determination. An infestation intensity of 20 BPH per hill and 5% tungro would therefore produce a loss of 6%.

The data from MRRTC yielded a different result. In this case, the influence of the brown planthopper was clearly greater than that of the virus infestation, although the absolute number of BPH per hill was not greater than in the corresponding infestation situation at BRES. Here the fact comes into play that the counting of brown planthoppers only once per season cannot give an adequate picture of the whole population development. It is nevertheless possible, using this infestation criterium as well as tungro infestation as further variables, to establish a statistically significant relationship with the dependent variable "yield loss".

A considerably higher precision of estimation was obtained by incorporating the multiplicative linkage of the two causes of damage as one variable. This indicates that the loss caused mainly by BPH is increased significantly by the interaction with the tungro infestation. The statistically significant intercept of 6.8 of the function again points to the existence of non-linear relationships in the range of low infestation intensities.

If we again assume an infestation intensity of 20 BPH per hill and 5% tungro, as in the case of the BRES function, and substitute these values in the MRRTC function, we obtain a crop loss of 12.8% as opposed to 6% at BRES.

This very clear difference between the two trial locations becomes smaller in the case of high infestation intensities due to the influence of the high Y intercept. These differences in the shape of the curve nevertheless show that location factors as well as infestation structure can considerably influence the results. For computing damage coefficients and deriving the corresponding threshold values for pests such as the brown planthopper, site-related trial analyses are therefore essential.

Overall, the results of the insecticide trials show that by means of the regression analysis derived from these trials it is possible, using the infestation levels of non-treated plots, to prepare differentiated loss estimates for non-resistant varieties. This considerably facilitates the analysis of the causes of crop loss which is important for economic analysis.

2.2.5.3 Results Obtained for Resistant Varieties Using Data from Field Trials

In computing damage coefficients for resistant varieties from the trials we had carried out during the period from 1979 to 1981, the main problem encountered was the low overall infestation intensity. To fulfill the statistical requirements, data from trials conducted at different locations and in different years had to be combined. This made it possible to construct separate functions for the major pests (Table 19). In the case of a number of pests, transformation of the variable "infestation" into logarithms considerably increased the coefficient of determination of the linear regression curve. The non-linear relationships existing particularly in the range of low infestation intensities are reflected better by this type of curve than by a non-transformed function. In the case of all pests, the negative sign of the *Y*-intercept points to the existance of a *physiological damage threshold*.

In the following, some explanations will be given regarding the functions listed in Table 19.

From the function for the rice caseworm, a *physiological damage threshold* corresponding to 4.4% infested leaves is obtained. An infestation intensity of 10% would result in a crop loss of 7.8%. From the function for defoliator, on the other hand, a physiological damage threshold of only 2.6% is obtained. An infestation of 10% thus causes a crop loss of 7.05%. The margin of error of the damage coefficient, the physiological damage thresholds and the confidence intervals of the regression parameters are given in Table 19. Since the number of observations is not very great, the confidence intervals are correspondingly large. In the case of the defoliator function, one additional percent of infestation can, for instance, result in a crop loss ranging from 0.78 to 1.136%. These extreme values of the confidence interval can be related to pessimistic or optimistic estimates of the yield-reducing effects of a specific intensity of infestation.

For the rice leaf folder, too, there is a *physiological damage threshold* beyond which each additional percent of infestation results in an increase in crop loss ranging from 0.467 to 1.957%. When using the lower limit of the confidence interval of the regression parameter and relating this to the mean value of the *Y*-intercept, the physiological damage threshold will be increased. The mean value of the latter in the case of an infestation during the generative growth stage amounts to 2.6% infested leaves.

In the trials conducted in the area of the pilot project for integrated pest management in Camarines Sur, the leaf folder was observed to occur jointly with defoliators. From these data it was possible to construct a function with two independent variables. The coefficients of the equation show that the rice leaf folder has a greater influence on yield than the defoliator complex, the incidence of the rice caseworm being negligible.

In the case of the rice bug, the results of the regression analysis based on field observations must be regarded as problematic since infestation intensity was measured in terms of adult insects whose activity and mobility vary greatly according to the time of day. This means that the magnitude of the infestation measurement (bugs m^{-2}) depends to a large degree on the subjective assessment of the observer. It therefore seems imperative to compare the function derived from field observation data with the results of the cage trials. According to the function derived from cage trials (Table 16), the physiological damage threshold corresponds to 0.36 bugs m^{-2}, beyond which each additional bug results in a loss of 2.52%. In view of the nature of the trial, exactness of infestation measurement can be assumed. Although the function constructed from the field data is largely consistent with the function derived from cage trial data, there is the possibility of infestation having been measured incorrectly in the case of the former.

To verify whether this function corresponds to reality, the control threshold based on empirical values which is currently used in Philippine rice farming, namely 2–4 bugs m^{-2}, can be used as a further criterium for comparison. The coefficients on which this threshold value is based can be computed – making realistic assumptions for control costs, control effectiveness and crop price – by resolving Norton's (1976) control threshold formula:

$$\Theta = \frac{C}{pdk}$$

where: Θ = infestation intensity at which control is worthwhile
C = control costs
p = price of pesticide
d = relationship between infestation intensity and crop loss, i.e. the loss coefficient
k = effectiveness of control in %

The following assumptions are made regarding the above factors:

cost of control:	120 Pesos ha^{-1}
effectiveness of control:	80%
price per kg:	1.00 Pesos kg^{-1}

The above assumptions correspond to the price-cost ratios at the time this threshold was first recommended. Since all the assumptions for the control thresholds currently used in Philippine rice farming are based on experimental data, corresponding data were also used for control costs and effectiveness of control. If the equation:

$$2 \text{ or } 4 \text{ respectively} = \frac{120}{1 \times 0.8 \times d}$$

is solved for d and the resulting value is applied to a yield level of 4 t ha^{-1}, then we obtain a damage coefficient such that each additional bug m^{-2} results in a loss of 1.88%. For a threshold of four bugs m^{-2}, we obtain a corresponding value of 0.94. The function in Table 20, on the other hand, results in an additional loss of 2.42% per unit of infestation on the basis of a *physiological damage threshold* of about four bugs per 10 m^2. Here it must be taken into account that the damage coefficient computed from field data is based on only two field trials and that the regression parameters therefore have high confidence intervals. This fact must always be borne in mind when computing the crop loss caused by this pest. In view of the fact that the damage coefficient for the rice bug is still rather uncertain, it would appear justified to continue using the one accepted to date.

Similarly the effect on yield of an infestation by the rice whorl maggot is still unclear. Up to 1979, the rice whorl maggot was still regarded as one of the major pests (Heinrichs et al. 1979). In recent studies carried out at IRRI in which various degrees of infestation intensities were simulated in the field, this insect was not found to have a yield-reducing effect, even in the case of high infestation intensities. It is nevertheless still considered possible that the *rice whorl maggot* might have a damaging effect if it occurs in conjunction with other pests.

In the case of this insect, quantitative loss assessment again poses a particular problem due to the fact that crop loss was estimated by means of a 1–9 scale of assessment (IRRI 1980). Even after conversion of the scale of values into percent, it was still impossible to derive a damage coefficient from the data obtained from field trials. For this reason, data from the individual plant trials were then used, including only such plants on which there had been minimal incidence of other pests. In order to prevent genetic influences having to be taken into account in the explanation of "yields per hill," the plants were grouped according to infestation intensity and the mean yield was then computed for the mean infestation intensity of the respective infestation intensity interval (Table 18). The *physiological damage threshold* was assumed to be 8% damaged leaves since hill yields varied greatly up to this level of infestation intensity and therefore no yield-affecting influence is expected.

The yield averages and infestation intensities were then converted into yield loss in percent and from these, the equation in Table 19 was derived. This function indicates an additional loss of 0.7% per percent of leaves infested by the *rice whorl maggot.* In accordance with the plausibility assumptions, the physiological damage threshold is roughly 8% infestation intensity. Although overall, this figure seems to contradict the more recent IRRI results mentioned above, however, the two results are not comparable. In the case of the IRRI figures, the damage was produced within a maximum of 30 days after transplanting and infestation in the plot field trials was assessed at roughly the same time (IRRI 1980).

Table 18. Average yields per hill at various infestation intensities of the Rice Whorl Maggot during the end of the vegetative growth stage

No. of plants	Infestation intensity intervals	Infest. intensity (%)	Yield (g per hill)	Yield loss (%)
17	0– 8	8	25.08	0.0
24	> 8–12	10	24.91	0.07
24	> 12–16	14	24.22	3.43
29	> 16–20	18	23.23	7.37
8	> 20–24	22	22.50	10.20
7	> 24–max	27	22.01	12.22

Source: Trials conducted by the author, Iloilo 1981–82, dry season

Table 19. Functions for estimating crop loss in resistant varieties

Pest	Basic data for computation of the function	Type of function regression parameter	Exogenous variables[a]	Confidence interval ($P \leq 0.05$)		n^b	R^2 F-value
				Y intercept	Regression coefficient		
1. Rice case worm	Average values of plot field trials	$y = -16.599 + 23.4479 \log (x+1)$	x = Percent infested leaves Maximum – veg. growth phase	$-26.56 < a_0 < -6.63$	$26.68 < b_1 < 42.681$	31	0.48 26.89
2. Defoliator	Completely treated and non-treated plots	$y = -2529 + 0.958x$	x = Percent infested leaves Maximum – veg. growth phase	$-4.8 < a_0 < -0.25$	$0.78 < b_1 < 1.136$	20	0.87 127.60
3. Rice leaf folder	Average values of plot field trials	$y = -37.01 + 34.68 \log (x+9)$	x = Percent infested leaves Maximum – gen. growth phase	$-48.01 < a_0 < -26.85$	$26.68 < b_1 < 42.68$	30	0.74 78.7
4. Rice leaf folder + defoliator	Completely treated and non-treated plots	$y = -0.544 + 0.7764x_1 + 0.8885x_2$	x_1 = Defoliator infestation Maximum – gen. growth phase n x_2 = Rice leaf folder Maximum – gen. growth phase		$0.4146 < b_1 < 1.138$ $0.5141 < b_2 < 1.2625$	21	0.87 61.73
5. Rice whorl maggot	Ind. plant trial, inf. groups (intervals of 4%)	$y = -4.17 + 0.709x$	x = Percent infested leaves towards end of veg. growth phase	$7.57 < a_0 < -1.01$	$0.355 < b_1 < 0.724$	6	0.98 178.6
6. Rice bug	Average values of plot field trials	$y = -1.01 + 2.4229x$	x = No. of bugs per m^2 at harvest time	$-6.98 < a_0 < 4.96$	$1.599 < b_1 < 3.24$	12	0.81 43.02

Table 19. Continued

Pest	Basic data for computation of the function	Type of function regression parameter	Exogenous variables[a]	Confidence interval ($P \leq 0.05$)		n^b	R^2
				Y intercept	Regression coefficient		
7. Stem borer + defoliator[c]	Ind. plant trial, all plants, 1981 rainy season	$y = -3.34 + 1.59x_1 + 0.75x_2 + 0.334x_3 - 0.16x_4 - 0.265x_5 - 0.227x_6$	x_1 = No. of panicles at harvest time	–	–	298	0.68 104.25
			x_2 = No. of tillers, gen. phase	–	–		
			x_3 = No. of tillers, veg. phase	–	–		
			x_4 = No. of white heads at harvest time	–	–		
			x_5 = No. of dead hearts, gen. phase	–	–		
			x_6 = Defoliator, gen. phase	–	–		
8. Sheath rot + stem borer, defoliator + rice whorl maggot	Ind. plant trial, all plants, 1981/82 dry season	$y = 6.706 + 0.973x_1 + 0.239x_2 - 0.0817x_3 - 0.0566x_4 - 0.1077x_5 - 0.251x_6 - 0.067x_7$	x_1 = No. of panicles at harvest time	–	–	254	0.58 48.65
			x_2 = No. of tillers, veg. phase	–	–		
			x_3 = Whorl maggot, gen. phase	–	–		
			x_4 = Sheat rot, % inf. tillers, ripening phase	–	–		
			x_5 = Defoliator, veg. phase	–	–		
			x_6 = White heads, ripening phase	–	–		
			x_7 = Sheat rot, % inf. leaves, veg. phase	–	–		

[a] The endogenous variable is always crop loss in percent
[b] No. of observations
[c] In this case the endogenous variable is yield in g/hill

In the individual plant trials, on the other hand, an infestation continuing well into the generative growth phase was observed. In all further computations of crop loss due to the rice whorl maggot, infestations occurring at an early plant growth stage were therefore neglected, whilst later infestations were taken into account by means of the above function.

In Table 19, two functions are listed (Nos. 7 + 8) by means of which, crop loss can be estimated for several jointly occurring pests. The basic disadvantage of function 1–6 is the fact that for these functions trials with the same major damage factor were combined whilst the remaining damage influences were not taken into account. Although these residual influences are only of minor importance due to the slight infestation levels involved, a slight influence on yield, especially on the untreated plot, cannot be ruled out. As a result, the damage coefficient of the main damage factor concerned may be overestimated. On the other hand, since nature and composition of this residual infestation vary from case to case, it is impossible to take into account the remaining factors by including one single additional variable.

A possibility of taking into account several jointly occurring pests was, however, provided by the data from the individual plant trials. In the statistical analysis of these data it had to be borne in mind that, strictly speaking, the data had not been obtained from individual plants but from hills. As a result, yield variance between hills was attributable primarily to the variables "number of panicles" and "tillers per hill". The corresponding results are also listed in Table 19. They show that in the 1981 rainy season, despite a low overall infestation intensity, *stem borer* as well as *defoliator* influence on hill yields was detected. The influence of other pests such as the *rice whorl maggot* or the *rice leaf folder,* on the other hand, was not found to be statistically significant.

For the statistically significant damage factors, we then computed the potential yield. This was done by calculating the averages of the yield-determining parameters, e.g. number of tillers and panicles, and substituting these values in the equation (No. 7) in Table 19. By multiplying the regression coefficients of the variables having a positive influence on yield, e.g. panicles and tillers, by their sample means, we thus obtained a potential yield of 45.97 g per hill for the function based on the 1981 rainy season data. If we now relate the values of the regression coefficients, i.e. 0.16 for stem borer white heads, 0.265 for stem borer dead hearts (generative growth phase) and 0.227 for defoliator to the computed potential yield, we obtain the following loss equivalents or damage coefficients: for white heads, a crop loss of 0.348% per percent of infestation, for dead hearts, a loss of 0.576% and for defoliator, a loss of 0.493%. Hereby the *physiological damage threshold* is neglected.

In the dry season, the result was totally different. There was a heavy sheath rot infestation [*Sarocladium oryzae* (Saw.) W. Gams & Hawksw.] with up to 70% unopened panicles on some of the hills. In addition, the *rice whorl maggot* occurred during the vegetative growth phase and infestation by this insect continued well into the generative phase. Fifty nine days after transplanting, 80% of the leaves were found to be infested on the most heavily infested hill. In addition, a few hills were infested by the *stem borer,* resulting in up to 37% white heads in the late growth stage and 59% dead hearts in the early growth stage.

Overall, taking into account the 300 individual plants, the influence of the stem borer was low, amounting to 1.2 and 1.95% respectively, so that the influence of *sheath rot* and *rice whorl maggot* dominated. Although the function constructed for the dry period has an unsatisfactory coefficient of determination of $R^2 = 0.58$, the influence of a total of four damage factors is statistically significant. Due to the influence of irrigation problems, the average potential yield amounting to 26.43 g per hill was clearly below the level of the preceding rainy season. The individual damage coefficients can be computed accordingly:

rice whorl maggot, gen. phase:	0.309
sheath rot, ripening phase:	0.214
defoliator, vegetative phase:	0.407
white heads, ripening phase:	0.949
sheath rot, vegetative phase:	0.225

Due to the different infestation structures, a direct comparison of the two functions (7 and 8 in Table 19) is impossible. It is only in the case of defoliator that an inference might be drawn to the effect that an infestation at a later growth stage (1981 rainy season) will result in a heavier crop loss than an earlier infestation (1981/82 dry season).

2.2.5.4 Summarizing Assessment of Findings and Comparison with the Relevant Extension Recommendations

The damage coefficients were computed for the currently used varieties IR-36, IR-42, IR-50 and IR-54. These varieties can practically be regarded as resistant to tungro and to the brown planthopper, the most dangerous pests. They are, however, susceptible to attack by pests which do not give rise to great epidemics, but produce crop losses of smaller magnitude which are more difficult to prove statistically. Strictly speaking, the available data do not meet the statistical requirements for the computation of damage coefficients, since they cover neither all the conceivable patterns of infestation development, i.e. early to late infestation, low and high intensities, nor all the possible combinations (Kelber 1975). Although in all cases trial data covering a period of at least 2 years were available, the influence of the trial site has not been taken into account. The parameters of the regression equations consequently have considerable confidence intervals so that there is a considerable margin of error in the *loss assessments* based on these parameters. It nevertheless seems justified to use the coefficients computed for further economic calculations, since the data available for the present study is more comprehensive than the data found to date in the relevant literature. Furthermore, mathematically derived relationships between infestation intensity and yield or crop loss, respectively, have the advantage of being more transparent and easier to reconstruct than pragmatic considerations and empirical values. Although the damage coefficients used to date in Philippine rice farming are also based on data from greenhouse and cage trials, the statistical proof of the existence of functional relationships between infestation intensity and crop loss was, with very few exceptions, dispensed with in these trials.

Although it can be assumed that loss assessments based on data obtained under controlled conditions will not be identical with *loss assessments* based on field trial data (Burleigh 1980), it is all the same necessary to check the latter data for its closeness to reality. Apparently, the parameter values determined in the field are lower. In the case of the *rice leaf folder,* for instance, the *physiological damage threshold* under greenhouse conditions is 7–8% infested leaves, while under field conditions it is computed to less than 3%. Furthermore, the regression coefficients for the other leaf pests seem to be unproportionately high (e.g. 0.958 for the *defoliator* complex and 0.539 for the *rice whorl maggot*) if the results of the trials with artificially produced leaf damage are used as a basis for comparison. It must, however, be taken into account that a single, non-recurring damaging of leaves cannot be compared with an ongoing infestation over a certain growth period of the crop. In the case of field trials, the effect on yield depends on the nature and intensity of pest population development so that an infestation assessment made during the critical development phase is a more or less true reflection of the population development concerned. This can be illustrated as follows: in the case of systematic infestation assessment – as in the case of *defoliator* – cumulated infestation intensity, up to about the heading stage, can be employed as an exogenous variable explaining crop loss. By means of the function:

$$Y = -3.37 + 0.3778\,x^{*} \qquad R^2 = 0.849,$$

where Y = crop loss in %,
x = cumulated defoliator infestation intensity,

the physiological damage threshold is computed as roughly 10% infested leaves, and a lower crop loss per additional unit of infestation is arrived at. For rice leaf folder plus defoliator, it was possible to proceed in the same way, since systematic measurements at regular intervals had been taken. In this case the function is given by:

$$Y = -1.728 + 0.303\,X^{**} + 0.562\,X_2^{***} \qquad R^2 = 0.914,$$

where Y = crop loss in %,
X_1 = cumulated defoliator infestation intensity, generative phase.

Although this function has a higher determination coefficient, the same overall tendency is obtained as in the case of defoliator.

From the above computations it can be concluded that the difference in the development of pest populations generally reduces the value of critical point models. More sophisticated models, on the other hand, do not only require more extensive data, but are also less practicable. Thus, if the area under the pest progress curve is to be used as a variable, it would mean that this area always has to be computed before any control decisions can be taken. If cumulated infestation intensity were to be used as a decision criterion, the farmers would either have to keep records or use other sampling aids.

Finally, the parameters computed will be compared with the control thresholds currently used (Table 20). For this purpose – as in the case of the rice bug – the damage coefficients were derived from the *control thresholds* currently used. The values used in these computations for control costs and effectiveness of control were based on the pesticides and dosage recommendations provided

Table 20. Comparison between various damage coefficients

1981 Position[a]			Author's own computations				
Pest	Control threshold	Damage coeff.[b]	Greenhouse trials		Field trials		Single plant method (damage coeff.)
			Phys. thresh.	Damage coeff.	Phys. thresh.	Damage coeff.	
Rice whorl maggot vegetative phase	75	0.0458	–	–	–	–	–
Rice whorl maggot generative phase	–	–	–	–	8.0	0.704	–
Rice caseworm	15	0.1875	–	–	4.10	1.8[e]	–
Defoliator vegetative phase	25	0.1125	–	–	2.60	0.958	0.407
Defoliator generative phase	15	0.1875	–	–	–	0.7764	0.493
Rice leaf folder vegetative phase	25	0.1875	–	–	–	–	–
Rice leaf folder generative phase	15	0.1325	7.3	1.82	– 2.7	0.885[d] 1.2[c]	–
Stem borer vegetative phase	15	0.24	5.89	0.775	–	–	–
Stem borer generative phase	5	0.875	2.6	1.634	2.77	1.4003[e]	0.576
Stem borer ripening phase	–	–	2.0	1.801	–	1.617	0.348[f] 0.949
Rice bug	2–4	1.88–0.94	0.36	2.52	0.41	2.42	–

[a] Source: Guidelines for a Surveillance and Early Warning System, Manila 1980, and author's computations up to 1981
[b] Obtained by means of Norton's formula for computing control thresholds
[c] Since the function is not linear, this figure only applies to the first unit of infestation above the physiological damage threshold
[d] In the case of several pests being involved
[e] Obtained from insecticide trials with varieties which were not tungro- and BPH-resistant
[f] Combined with infestation during the generative phase

by the extension service and the kill rate generally achieved under trial conditions. Strictly speaking, the figures given in Table 20 in the column "1981 position" cannot be regarded as damage coefficients since they are not the result of empirical investigations. However, use of the formula developed by Norton (1976) automatically results in these orders of magnitude, provided the other assumptions on which the current control thresholds are based are correct.

Compared with the author's computations, the crop loss per unit of infestation intensity computed from the current damage coefficients seems to be low. This is due to the fact that – as opposed to the "single-pest" approach customary to date – the author has in his computations generally taken into account several pests. Especially in the case of leaf pests, the threshold values recommended to date seem to be too high.

For the *stem borer,* clear results cannot be arrived at. According to the computations based on the insecticide trials, the damage coefficients should be higher. Although to a certain extent it would be possible to use these values for the current varieties which possess only a medium stem borer resistance, it seems doubtful whether the coefficients computed for a generally high level of stem borer infestation are suitable for the current conditions of a low level. On the other hand, studying the results of the individual plant trials, it will be noted that the value arrived at for combined infestation during the generative and ripening phases, namely $0.576 + 0.348 = 0.924$, is similar to the presently used *control threshold* at the generative phase (0.875).

Summing up, it is clear that the parameters computed are not entirely satisfactory from a scientific point of view. They do, however, permit further economic calculations to be carried out covering the major infestation situations in resistant and non-resistant varieties.

2.2.6 Computation of Crop Losses on the Basis of Infestation Data and Damage Coefficients

The computation of crop loss by means of a simple comparison of the yields from treated and untreated plots has been dealt with in Section 2.1. The usefulness of the results obtained by this method, which are often found inadequate, can be enhanced if statistically significant damage coefficients are available. Crop loss can then be computed from the infestation data by substituting the latter in the respective loss functions. In this way, even a small crop loss, which in a simple comparison of yields would perhaps not be statistically significant, can be computed relatively exactly. A sizeable further advantage of this method is that, in contrast to simple yield comparisons, it also permits the computation of loss profiles, i.e. of the contributions of the individual pests to overall loss.

The most important data required for this purpose are infestation levels from trial plots or observation fields which should, if possible, be entirely free of the effects of insecticides. If it is intended to compute crop loss through time, data from different cultivation areas, covering several years must be available.

In the following computations it was again possible, for non-resistant varieties, to fall back on infestation data from the official pesticide evaluation trials. For resistant varieties, it was possible to use infestation measurements from the field trials conducted to determine alternative control methods, along with data from the author's own trials. In addition, SEWS infestation monitoring data covering several years were available from one observation station each in the Iloilo and Camerines Sur cultivation areas. However, the latter data do not truly reflect the actual infestation situation, since they were influenced by pesticide applications.

To begin with, the entire data must be checked for any differences in infestation intensities between different sites and between rainy and dry seasons. On the basis of the result it can be determined to what extent it is necessary for the computations of crop loss to be differentiated.

2.2.6.1 Analysis of Pest Development Over Time

Only scant data are available on the regional distribution of the different pests. Cendena and Calora (1964) point out that some insect pests depend to a certain extent on a specific type of climate. The rice bug, for instance, occurs primarily in areas with evenly distributed rainfall.

In the case of the brown planthopper, too, many authors (Dyck 1978) assume that drought encourages the mass reproduction of this insect. In areas with evenly distributed rainfall, the probability of an outbreak is consequently not as great. For the tungro virus, on the other hand, these same areas provide particularly favourable preconditions for epidemics.

Apart from climatic factors, a number of other factors can be associated with pest outbreaks (Christ and Lever 1969). The interaction of these complex correlations has to date not been studied in depth. For this reason, it will suffice at this point to compare the different infestation structures in the individual investigation areas without going into any detail as regards their causes.

2.2.6.1.1 Non-Resistant Varieties. The counts taken on untreated plots in pesticide evaluation trials in which the IR-22 variety was used can be tested for differences between years and between cultivation periods. Table 21 (a–c) lists the average annual infestation intensities for the three experiment stations VRES (Iloilo), BRES (Camarines Sur) and MRRTC (Nueva Ecija). Differences between years could not be detected by means of the analysis of variance.

A comparison between dry and rainy seasons, on the other hand, yielded significant differences in the *t-test* in the case of some of the pests.

At VRES, this applies to the *brown planthopper:* in the rainy season – despite a low overall infestation pressure – incidence is clearly higher than during the dry season. This observation is also confirmed at the BRES station, but does not apply to MRRTC. If we take into account the different geographical locations of the experimental stations, this result contradicts the view referred to above, which is held by a number of authors. However, VRES and BRES results are also confirmed by an Indonesian study. According to this study there is a clearly positive correlation ($r = 0.907$) between the total annual rainfall and the size of the infested area (Sastrosoedrjo, 1979). The number of rainy days per year used as explanatory variable, also resulted in a correlation of 0.838, although – in view of the shortness of the period concerned (1970–1975) – the value of this result should not be overestimated.

No significant differences in infestation intensity were found to exist in the case of the tungro virus, the highest figure of over 30% in both the rainy and the dry season being measured at BRES.

For the stem borer, different results are obtained at the different stations and during the different infestation periods. Although at VRES and MRRTC more dead hearts and white panicles are detected during the dry season, it is only in the case of MRRTC and only for an infestation during the vegetative period, that a statistically significant difference at $P \leq 0.1$ can be found. At BRES, the reverse is the case, the differences in generative phase of the dead hearts and white panicles being found to be statistically significant. It must, however, be

Table 21a. Infestation intensities in untreated plots of insecticide trials (VRES, 1970–1980)

Year	Pest/Growth phase																	
	Stem borer[a]						Leaf folder[b]		Brown planthopper[c]				Green leafhopper[d]				Tungro[e]	
	dh 30	*n*	dh 50	*n*	wh	*n*		*n*	Veg.	*n*	Gen.	*n*	Veg.	*n*	Gen.	*n*	Veg.	*n*
1970	3.34	5	2.21	5	4.22	5	–	–	–	–	–	–	–	–	–	–	–	–
1971	6.7	2	4.97	2	4.68	2	–	–	–	–	–	–	–	–	–	–	–	–
1972	1.16	2	3.45	2	1.92	2	3.0	1	5.0	1	–	–	–	–	–	–	–	–
1973	–	–	–	–	–	–	–	–	–	–	–	–	–	–	–	–	–	–
1974	3.58	3	2.12	3	1.74	3	–	–	–	–	–	–	–	–	–	–	–	–
1975	0.63	1	1.36	2	3.06	2	–	–	11.0	1	11.0	1	20.5	2	21.25	2	4.0	1
1976	8.89	4	12.6	4	10.82	4	4.05	4	24.84	5	12.24	5	7.92	5	5.15	4	9.4	1
1977	5.86	5	7.29	5	10.27	5	–	–	4.45	2	2.85	3	34.47	3	2.95	2	2.0	1
1978	5.08	7	6.88	6	5.85	6	21.61	1	9.61	6	4.63	5	9.91	5	11.38	5	1.1	3
1979	0.74	8	1.16	8	1.35	7	2.41	2	3.37	4	8.17	3	9.07	6	18.03	4	–	–
1980	0.56	4	0.71	4	0.95	4	–	–	12.41	4	38.33	3	5.03	4	8.57	4	–	–
Average ($\bar{x}$)	3.71		4.32		4.72		5.71		11.74		12.17		12.3		11.06		3.11	
$\bar{x}$ – rainy season	3.54		4.19		4.55		–		13.59		16.3		14.23		13.3		–	
$\bar{x}$ – dry season	4.33		4.87		5.28		–		5.06		2.52		7.33		5.46		–	
Difference acc. to *t*-test	$0.79^{n.s.}$		$0.68^{n.s.}$		$0.73^{n.s.}$		–		8.53*		13.78**		$6.9^{n.s.}$		0.018*		–	

Significance level acc. to *t*-test: * $P \leq 0.1$
** $P \leq 0.05$
*** $P \leq 0.01$

[a] Percentage of dead hearts (dh) or white heads (wh)
[b] Infested leaves
[c] No. per hill
[d] No. per sweep
[e] Percentage of infested hills

n.s. = Not significant
n = No. of trials
– = No data available

Table 21b. Infestation intensities in untreated plots of insecticide trials (BRES, 1972–1981)

Year	Pest/Growth phase																			
	Stem borer[a]						Leaf folder[b]		Brown planthopper[c]				Green leafhopper[d]				Tungro[e]			
	dh 30	*n*	dh 50	*n*	wh	*n*		*n*	Veg.	*n*	Gen.	*n*	Veg.	*n*	Gen.	*n*	Veg.	*n*	Gen.	*n*
1972	2.1	1	3.6	1	–	–	–	–	–	–	–	–	–	–	–	–	–	–	43.0	1
1973	2.08	2	3.94	3	2.76	4	14.8	1	–	–	–	–	177.5	2	–	–	4.8	1	10.95	4
1974	–	–	–	–	–	–	–	–	–	–	–	–	–	–	–	–	–	–	–	–
1975	2.92	1	2.33	1	2.36	1	–	–	–	–	–	–	163.0	1	560.0	1	16.7	1	17.3	1
1976	–	–	–	–	–	–	–	–	–	–	–	–	–	–	–	–	–	–	–	–
1977	2.3	5	2.62	5	2.81	8	–	–	75.16	6	169.6	3	197.4	7	283.8	5	12.04	8	47.17	3
1978	0.93	6	1.49	12	1.01	12	8.3	1	155.5	10	250.03	11	286.0	10	363.8	12	3.5	5	14.33	13
1979	0.72	4	2.0	5	1.79	5	–	–	130.0	4	262.0	5	270.0	5	314.0	5	7.88	5	44.81	6
1980	0.55	10	1.46	13	0.38	12	–	–	176.9	13	286.7	13	291.5	13	307.0	13	4.71	13	36.66	13
1981	0.41	4	0.53	5	0.1	5	–	–	42.0	5	80.0	5	258.0	5	272.0	5	8.58	5	55.24	5
Average ($\bar{x}$)	1.1		1.79		1.31		11.5		138.13		234.8		260.0		323.0		7.33		31.60	
$\bar{x}$ – rainy season	1.05		2.059		1.63		–		181.29		331.92		256.8		339.8		6.79		32.03	
$\bar{x}$ – dry season	1.25		1.24		0.69		–		48.92		92.4		266.6		295.3		8.39		30.86	
Difference acc. to *t*-test	$0.20^{n.s.}$		0.819*		0.94**		–		132.37***		238.6***		$9.8^{n.s.}$		$44.5^{n.s.}$		$1.6^{n.s.}$		$1.17^{n.s.}$	

Significance level acc. to *t*-test: * $P \leq 0.1$
** $P \leq 0.05$
*** $P \leq 0.01$

[a] Percentage of dead hearts (dh) or white heads (wh)
[b] Infested leaves
[c] No. per hill
[d] No. per sweep
[e] Percentage of infested hills
n.s. = Not significant
n = No. of trials
– = No data available

Table 21c. Infestation intensities in untreated plots of insecticide trials (MRRTC, 1972–1981)

Year	Pest/Growth phase																	
	Stem borer[a]						Brown planthopper[b]				Green leafhopper[c]				Tungro[d]			
	dh 30	*n*	dh 50	*n*	wh	*n*	Veg.	*n*	Gen.	*n*	Veg.	*n*	Gen.	*n*	Veg.	*n*	Gen.	*n*
1967	–	–	4.04	1	11.5	1	–	–	–	–	–	–	–	–	–	–	–	–
1968	–	–	7.24	1	8.6	1	–	–	–	–	–	–	–	–	–	–	–	–
1969	–	–	12.3	1	1.1	1	–	–	–	–	–	–	–	–	–	–	–	–
1970	9.0	2	14.6	2	8.3	2	–	–	–	–	–	–	–	–	–	–	–	–
1971	–	–	–	–	–	–	–	–	–	–	–	–	–	–	–	–	–	–
1972	–	–	–	–	–	–	–	–	–	–	–	–	–	–	–	–	–	–
1973	15.3	1	2.49	5	5.14	5	–	–	–	–	–	–	–	–	–	–	–	–
1974	8.94	1	11.76	1	17.12	1	–	–	–	–	–	–	–	–	–	–	9.8	2
1975	4.8	1	4.62	2	2.91	3	22.0	1	–	–	–	–	–	–	–	–	–	–
1976	–	–	0.06	2	0.03	2	1.01	2	–	–	–	–	–	–	–	–	49.75	2
1977	2.19	2	1.37	4	1.14	4	6.86	2	72.27	3	17.2	1	–	–	–	–	86.52	2
1978	–	–	0.62	1	–	–	–	–	–	–	31.96	3	69.5	2	4.9	2	7.03	4
1979	0.83	8	0.62	8	0.84	8	3.1	2	7.48	5	–	–	–	–	0.01	1	36.3	1
1980	0.98	13	0.92	13	0.81	13	10.69	6	25.88	13	17.33	8	19.74	7	0.69	6	0.96	2
1981	0.22	3	0.26	3	3.54	2	14.17	3	22.15	3	22.85	13	30.55	13	0.17	8	2.89	13
											19.5	2	54.0	2	1.15	3	26.77	3
Average ($\bar{x}$)	2.31		2.5		2.75		9.41		27.38		21.77		32.6		0.94		16.43	
$\bar{x}$ – rainy season	1.37		1.87		2.68		9.36		29.91		26.37		32.73		–		18.47	
$\bar{x}$ – dry season	4.0		3.61		2.85		9.55		21.23		12.59		32.33		–		12.53	
Difference acc. to *t*-test	2.63*		1.74[n.s.]		0.17[n.s.]		0.19[n.s.]		8.68[n.s.]		13.78**		0.4[n.s.]		–		5.94[n.s.]	

Significance level acc. to *t*-test: * $P \leq 0.1$, ** $P \leq 0.05$, *** $P \leq 0.01$

[a] Percentage of dead hearts (dh) or white heads (wh)
[b] No. per hill
[c] No. per sweep
[d] Percentage of infested hills
n.s. = Not significant
n = No. of trials
– = No data available

noted that within the overall infestation at this station, the stem borer is only of minor importance. It is clear, on the other hand, that the results call for a differentiation of the data according to rainy and dry season. For this reason the computation of crop loss has been based on the average infestation intensities per year and per cropping period.

2.2.6.1.2 Resistant Varieties – Untreated Plots. For the Provinces of Iloilo and Nueva Ecija, data on infestation intensities on untreated plots are available from 1976 and 1978 onwards respectively; for Camarines Sur, on the other hand, data exist only from three field trials. Since in contrast to the non-resistant varieties, the infestation levels encountered in the resistant varieties were low, it could not be assumed from the beginning that significant differences would be found. To start with, the data from the different production sites were therefore compared. Table 22 gives the minimum, maximum and average annual infestation intensities. However, due to the lack of homogeneity of the variances within years, a statistical comparison by means of the analysis of variance was not possible. However, it is nevertheless possible to detect differences in the infestation structure at the three locations. In Iloilo, the incidence of *rice caseworm* and *rice bug* can be regarded as exceptionally high. In Nueva Ecija, the *rice leaf folder* seems to be gaining greater importance, whilst in Camarines Sur, the incidence of *defoliator* and *rice whorl maggot* is increasing. For our purposes, a comparison of the infestation intensities measured at one and the same site in different years does not seem very useful, since in many cases only a few observations per year are available. In the case of rainy and dry seasons, too, significant differences could be detected in only a few cases, as shown in Table 23.

In contrast to non-resistant varieties, infestation levels in resistant varieties tend to be higher in the dry season than in the rainy season. However, due to the low overall infestation levels and the limited number of observations available, statistically significant differences cannot be ascertained. From this point of view, a break-down of the data according to rainy and dry season does not seem necessary for identifying the optimum control strategy. On the contrary, it can be assumed that the infestation situations observed in the different trials are just as likely to occur in the rainy as in the dry season.

The economic analysis, on the other hand, should definitely be prepared separately for the different production sites. In the case of Iloilo and Nueva Ecija – due to the greater number of observations available – the obvious predominance of specific pests at the different sites can be shown to be statistically significant. By means of the *t-test,* a significant difference was ascertained for *rice caseworm* and *defoliator* at $P = 0.1$. For *stem borter – white head, rice leaf folder* and *rice bug,* the existence of a difference between means was ascertained with a probability of error of $P = 0.05$.

2.2.6.1.3 Resistant Varieties – Observation Stations. For the purpose of crop loss assessment, the figures available from the systematic pest monitoring of the official warning service can only be used with certain reservations. As infestation measurements are made on specific fields of farmers, the counts taken are affected by the chemicals applied by the farmers. Therefore for comparing the

Table 22. Range and average in infestation intensities in untreated trial plots, resistant varieties, three experimental stations[a]

Location	Pest									
	Rice case-worm	Defoliator		Rice whorl maggot		Stem borer			Rice leaf folder	Rice bug
		Veg. phase	Gen. phase	Veg. phase	Gen. phase	Veg. phase	Gen. phase	Ripening phase		
Iloilo										
$\bar{x}$[b]	8.21	0.42	0.54	6.67	0.25	1.61	0.95	2.37	3.01	3.55
Range[c]	0–54.0	0–26.0	0–4.15	0–65.6	0–4.0	0–6.1	0–5.5	0–7.4	0–13.0	0–14.0
Nueva Ecija										
$\bar{x}$	0.22	5.94	0.39	8.65	0.18	2.43	3.34	4.75	12.3	0
Range	0–2.8	0–28.9	0–5.1	0–40.0	0–2.4	0–7.7	0–12.5	0–11.5	0–35.0	0–0
Camarines Sur										
$\bar{x}$	0.17	4.91	12.8	33.14	12.23	4.77	5.36	4.91	6.2	0.72
Range	0–0.52	1.18–10.4	2.44–20.5	21.85–45.9	6.58–16.7	0.01–10.6	0.1–9.3	0–9.7	0–17.0	0–2.17

[a] For sampling units used (see footnote, Table 21)
[b] Arithmetic mean
[c] Lowest and highest values measured
Source: IRRI Insecticide Evaluation Report 1979–1981 and author's trials

Table 23. Differences in pest infestations in untreated plots of field trials between rainy and dry season, resistant varieties[a,b]

Location	Iloilo			Nueva Ecija			Camarines Sur			Total		
Pest	Rainy season	Dry season	Difference acc. to t-test	Rainy season	Dry season	Difference acc. to t-test	Rainy season	Dry season	Difference acc. to t-test	Rainy season	Dry season	Difference acc. to t-test
Rice leaf roller	14.59	0.0	14.59^{-}	0.35	0.0	$0.35^{n.s.}$	0	0.26	-0.26^{-}	7.45	0.03	7.42*
Defoliator veg. phase	0.29	0.58	$-0.29^{n.s.}$	8.05	2.57	$5.48^{n.s.}$	1.18	6.77	-5.59^{-}	3.78	2.17	$1.61^{n.s.}$
Defoliator gen. phase	0.46	0.63	$-0.17^{n.s.}$	0.64	0.0	0.64	2.44	17.98	-15.55^{-}	0.65	2.89	$-2.24^{n.s.}$
Rice whorl maggot veg. phase	7.43	5.72	$1.71^{n.s.}$	13.8	0.4	13.2**	42.97	56.46	-13.49^{-}	12.23	7.03	$5.2^{n.s.}$
Rice whorl maggot gen. phase	0.0	0.58	-0.58^{-}	0.3	0.0	$0.3^{n.s.}$	6.48	15.11	-8.63^{-}	0.49	2.45	$-1.96^{n.s.}$
Stem borer veg. phase	1.08	2.22	$-1.14^{n.s.}$	1.92	3.44	$-1.52^{n.s.}$	3.72	5.33	-1.61^{-}	1.63	3.06	$-1.43^{n.s.}$
Stem borer gen. phase	0.84	1.1	$-0.26^{n.s.}$	2.49	5.03	$2.51^{n.s.}$	6.69	4.69	2.0^{-}	1.97	3.0	$-1.03^{n.s.}$
Stem borer ripening phase	2.51	2.21	$0.3^{n.s.}$	4.27	5.5	$1.23^{n.s.}$	5.08	4.84	0.24^{-}	3.44	3.76	$-0.32^{n.s.}$
Leaf folder	3.51	2.38	$1.13^{n.s.}$	13.38	10.56	$2.82^{n.s.}$	1.59	8.52	-6.93^{-}	7.78	6.18	$1.61^{n.s.}$
Rice bug	2.6	4.78	$-2.18^{n.s.}$	0.0	0.0	0.0	0.0	1.09	-1.09^{-}	1.3	2.55	$-1.22^{n.s.}$

[a] Level of significance according to t-test:
* $P \leq 0.1$; ** $P \leq 0.05$; n.s. = not significant; – = significance level not computed

[b] For sampling units used (cf. Table 21)

Source: Author's own computations based on the data presented in Table 21

Table 24. Comparison of pest infestation in two SEWS observation stations, Iloilo and Camarines Sur, 1976–1981[a,b]

Location	Pest									
	Rice case-worm	Defoliator		Rice whorl maggot		Stem borer			Rice leaf folder	Rice bug
		Veg. phase	Gen. phase	Veg. phase	Gen. phase	Veg. phase	Gen. phase	Ripening phase		
Iloilo	0.25	1.71	1.27	6.75	4.02	2.45	2.60	1.4	8.98	0.69
Camarines Sur	0.05	0.57	1.09	16.29	11.44	1.18	2.40	2.36	0.99	3.7
Difference acc. to *t*-test	0.2064[n.s.]	1.14[n.s.]	0.18[n.s.]	−9.54**	7.42**	1.27[n.s.]	0.2[n.s.]	−0.96[n.s.]	7.99**	3.01***

[a] Significance level acc. to *t*-test:
*** $P \leq 0.01$; ** $P \leq 0.05$; n.s. = not significant
[b] For sampling units used (cf. Table 21)
Source: Infestation levels measured at observation stations, Bureau of Plant Industry Naga City and Iloilo City; unpublished data

Table 25. Difference in pest infestation in rainy and dry seasons in two SEWS observation stations[a]

Pest	Area								
	Iloilo			Camarines Sur			Total		
	Rainy season	Dry season	*t*-Test	Rainy season	Dry season	*t*-Test	Rainy season	Dry season	*t*-Test
Rice caseworm	0.02	0.41	$0.39^{n.s.}$	0.08	0	0.08^{-}	0.05	0.25	$0.2^{n.s.}$
Defoliator, veg. phase	1.7	1.72	$0.02^{n.s.}$	0.35	0.85	$0.5^{n.s.}$	0.95	1.33	$0.38^{n.s.}$
Defoliator, gen. phase	1.19	1.34	$0.15^{n.s.}$	0.52	1.8	1.28	1.88	1.52	$0.36^{n.s.}$
Rice whorl maggot, veg. phase	7.33	6.28	$1.05^{n.s.}$	14.43	18.6	4.17	10.88	11.21	$0.33^{n.s.}$
Rice whorl maggot, gen. phase	1.45	6.17	$4.72^{n.s.}$	6.10	18.10	$12.0^{n.s.}$	3.78	10.93	7.15*
Stem borer, veg. phase	1.87	2.83	$0.96^{n.s.}$	2.05	0.09	$1.96^{n.s.}$	1.97	1.73	$0.24^{n.s.}$
Stem borer, gen. phase	2.74	2.49	$0.25^{n.s.}$	2.55	2.22	$0.33^{n.s.}$	2.65	2.38	$0.27^{n.s.}$
Stem borer, ripening phase	1.38	1.42	$0.04^{n.s.}$	2.87	1.84	$1.03^{n.s.}$	1.97	1.61	$0.36^{n.s.}$
Leaf folder	10.97	6.99	$3.98^{n.s.}$	0.39	1.76	$1.37^{n.s.}$	6.16	4.9	$1.26^{n.s.}$
Rice bug	0.83	0.52	$0.31^{n.s.}$	4.6	2.58	$2.02^{n.s.}$	2.54	1.44	$1.1^{n.s.}$

[a] Significance level (*t*-test) between rainy and dry season:
* $P<0.1$; n.s. = not significant; – = test not carried out

Source: See Table 24

data from two observation stations, it must either be assumed that the effectiveness was identical or else must be quantified. If it is known, then the infestation intensity in the case of non-control can be roughly estimated from the measured infestation intensity. However, since the data concerned are time series data, it is impossible for the effectiveness achieved in an individual case to be determined subsequently. For the analysis of infestation structure, conversion of the data will therefore be dispensed with. The data concerned are the counts taken at the two observation stations set up around 1976, one each in the vicinity of the pilot projects in Iloilo and Camarines Sur.

When the pilot projects for the introduction of integrated pest management were launched in 1980, these observation stations were discontinued and replaced by numerous assessment plots on farmers' fields. To enable the infestation time series of the observation fields to be continued beyond 1980, one of the test farms was always selected on a random basis.

The data collected by the warning service for the Nueva Ecija pilot project were very incomplete and therefore not used. The statistical comparison between Iloilo and Camarines Sur is shown in Table 24, in which the incidence of the rice whorl maggot is significantly higher in the Camarines Sur pilot project than in Iloilo, thus confirming the information derived from Table 22. On the other hand, in Iloilo – in contrast to the results from the untreated plots – leaf folder incidence is higher and rice bug incidence significantly lower than in Camarines Sur. The latter must again be viewed in connection with the small size of the sample from untreated plots in Camarines Sur. Comparing, as above, the rainy and dry season figures, we arrive at a result similar to that in the case of the untreated plots (Table 25). For this reason, a break-down of the data according to rainy and dry season is dispensed with in Table 24.

2.2.6.2 Results of Crop Loss Computations

Crop losses have been computed separately for resistant and non-resistant varieties and for the different locations. The computations are based on the infestation data given in Tables 21 to 24 and on the functions listed in Tables 17 and 19.

In the computations for non-resistant varieties, functions related to the respective trial location were used. It was only in the case of VRES that it was found impossible to estimate a function for tungro and for the brown planthopper. The corresponding figures from this station were consequently substituted in the respective functions derived from BRES data.

Since it was found impossible, in the case of the resistant varieties, to construct separate functions for the different locations, identical damage coefficients had to be assumed for the various pests at the different trial locations. The crop loss relationships generally refer to individual pests and it was only by analysing the individual plant trials that it was possible to construct multiple regression equations. When several pests occurred simultaneously, it was therefore impossible to encode all the existing combinations in corresponding functions. The procedure adopted was to use the function coming closest to the infestation combination concerned.

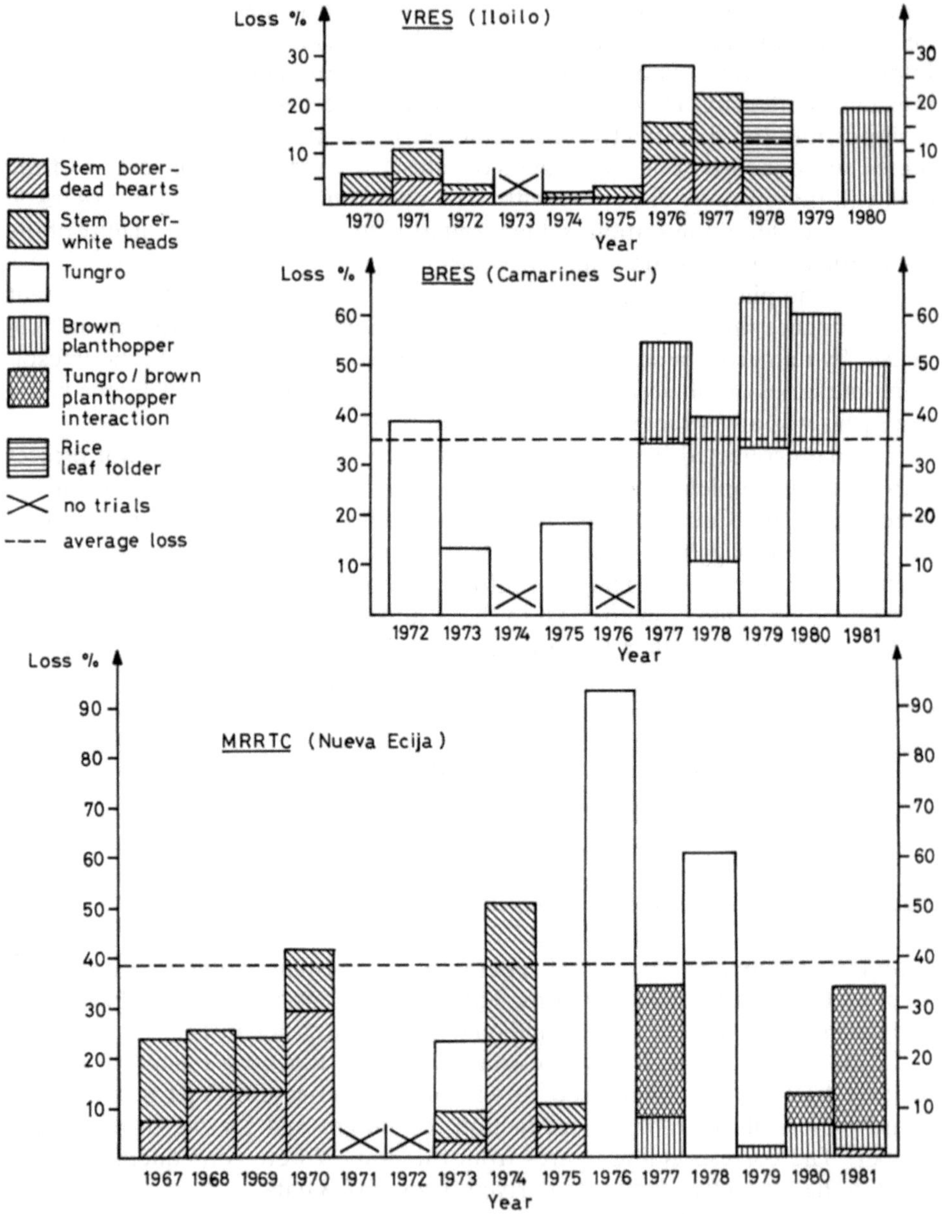

Fig. 4. Loss due to insect pest and virus diseases in non-resistant varieties in untreated plots at three experiment stations, 1976–1981

2.2.6.2.1 Non-Resistant Varieties. Figure 4 shows the computed crop losses and their time and location-dependent differences, crop losses being expressed in terms of average values. Average crop losses for VRES amount to about 12%, for Camarines Sur to 35% and for MRRTC to almost 40%. Comparing these figures with the average yield differences listed in Table 6, we find that for the BRES and Iloilo stations, the data are largely consistent, while for MRRTC, the losses computed by means of damage coefficients are 10% higher. This is due to the high percentage of non-significant yield differences which, when means are compared, are denoted by zero, but which become greater than zero when the infestation data are substituted in the corresponding functions.

Figure 4 also shows the development in time of the infestation/crop loss structure at the different trial stations. From the end of the 1960's up to the beginning of the 1970's, the loss profile of BRES and MRRTC is determined exclusively by the stem borer. However, overall losses were lower, amounting to less than 10% in VRES and generally to about 20% at MRRTC. From about the middle of the 1970's onwards, the main damage factors changed, especially at the MRRTC. It is interesting to note that the brown planthopper now often occurs jointly with tungro and that crop loss is subject to great annual fluctuation. Here a certain influence of factors favourable or unfavourable for tungro development can be assumed to exist, the two main determinants being availability of food for the vector and time. As already mentioned, climatic conditions in the area studied are such that there is an extremely pronounced difference between rainy and dry seasons. Rainfall may all the same occasionally occur during the dry season, resulting in a regrowth of the harvested rice plants (ratooning) and consequently in more favourable reproduction conditions for the green plant hopper. However, maximum crop loss amounting to over 90% is experienced when tungro occurs alone.

In the BRES area, the climatic preconditions necessary for tungro epidemics virtually always exist, although here, too, crop loss fluctuates. However, the increased incidence of the brown planthopper since about 1977 cannot be attributed to climatic factors alone, and it is considered possible that it is due primarily to the stimulating influence of certain chemicals (Chelliah 1979). Since 1979 there has been a slight downtrend in the contribution of the brown planthopper to the overall crop loss. Joint occurrence of brown planthopper and virus generally results in crop loss of about 50%, which can reach a maximum of about 65%.

For detailed economic analysis, it seems expedient, in view of the change taking place in the loss profiles of MRRTC and BRES, to use only infestation data from 1976 onwards. In this way, account is taken of the decreasing importance of the *stem borer*. Although the latter does not apply to VRES, the method nevertheless seems justified since loss diversification at this station also started at roughly this time.

In this way, the basic assumptions for the loss computations and the subsequent economic calculations based on the latter can be adapted to current conditions despite the limited number of "infestation events."

2.2.6.2.2 Resistant Varieties – Untreated Plots. Data for non-resistant varieties on untreated plots are available only from Iloilo and Nueva Ecija. For a loss assessment on the basis of these data, we can use the functions given in Table 19. In Iloilo, crop loss in some cultivation periods is determined to a large degree by a high *rice bug* incidence. Using the function constructed for the rice bug, sizable losses are arrived at. As already pointed out, quantification of the damage co-

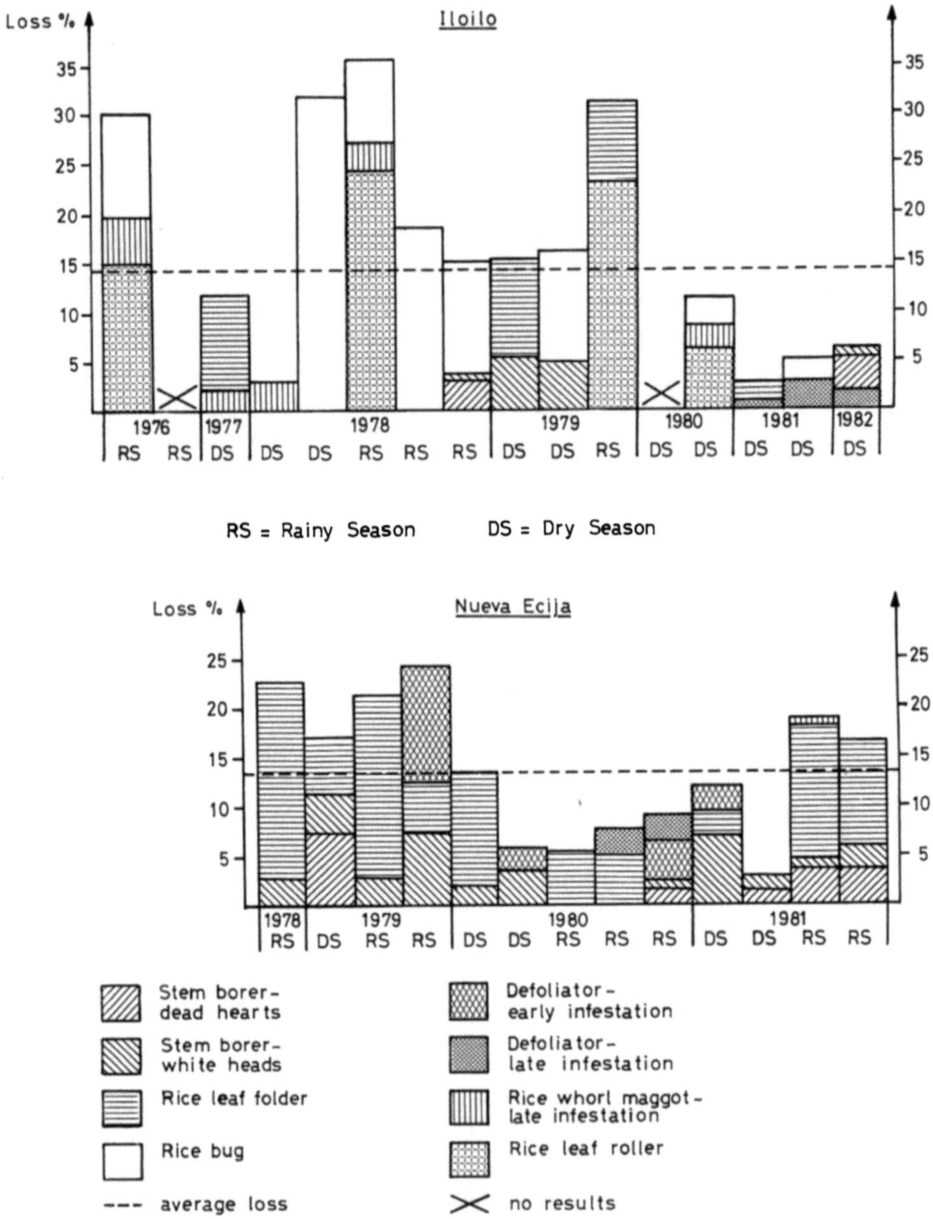

Fig. 5. Loss due to insect pests in resistant varieties in untreated plots of field trials, two investigation areas, 1976–1981

efficient for the rice bug is particularly problematic. It is therefore advisable to employ alternative methods of computation using the lower values of the confidence interval of the function. The control threshold for this insect was changed several times when it was found that the original threshold, namely two to four bugs per m^2, did not result in any significant additional yields. In a number of test trials carried out by Litsinger et al. (1980), the *control threshold* was four bugs per m^2 during the period from 1975 to 1978 and rose to six bugs per m^2 in 1979. An increase of this magnitude cannot be attributed to economic factors (e.g. a corresponding increase in the cost of pesticides). It is more likely that the damage caused by this insect had previously been overestimated. According to the result obtained by substituting the rice bug infestation data in the corresponding function (cf. Table 19), this pest would on the average account for about 50% of overall crop loss. On the average, taking into account all infestation events, the contribution of this pest would therefore amount to almost 15%. On the other hand, using the lower limit of the confidence interval of the function parameters, infestation would exceed the physiological damage threshold in only four cases, and crop loss would on average amount to 8.85%. A corresponding comparison with Table 8 shows that crop loss computed by means of this function is higher than the loss obtained by comparing the yields of treated and untreated trial plots. This fact is again due to the high proportion of non-significant trial results.

Converting the infestation data from the corresponding trials in Nueva Ecija into loss, we find that this amounts to 13% on average, the *rice leaf folder* and the *defoliator* complex being the dominant cause of loss (Fig. 5). In 6 out of 13 cases, these damage factors accounted for crop loss of over 10%. In five cases, on the other hand, the overall loss per trial was lower than 10%. In the case of the data from Nueva Ecija, this is also the main cause explaining the divergence from the data given in Table 8. In this table, average crop loss of almost 10% is listed for the rainy season, while for the dry season, they amount to less than 2%. The latter is mainly due to the lack of effectiveness of the maximum treatment as a result of which the potential yield – i.e. the basis for computing the relative yield difference – is definitely underestimated.

The loss caused by the stem borer always constitutes a residual value, exceeding 10% in only one case (Fig. 5).

2.2.6.2.3 Resistant Varieties – Observation Stations. Since these infestation data can be assumed to have been influenced by chemical control measures, the crop loss computed on their basis is too low.

The situation at the observation stations selected in Iloilo clearly indicates that crop loss has been decreasing since 1976 (Fig. 6). The loss profile, also has changed in as much as a great diversification is taking place. Average crop losses are of the order of 8.5%.

Although there were no continuous data series for the Camerines Sur observation station, the loss structure is quite evident. As in the case of the data from untreated plots in Iloilo, the results obtained for the rice bug are not clear. By substituting the corresponding infestation data in the function in Table 19, losses amounting to over 10% would repeatedly be obtained for this pest. On the other hand, if the lower limit value of the regression parameters is used as before, the average loss would drop from 13.6% to 6.5%. If we disregard the rice bug altogether, changes again take place in the composition of the infestation with the rice whorl maggot the incidence of which continues right into the generative growth phase now accounting for the major part of it.

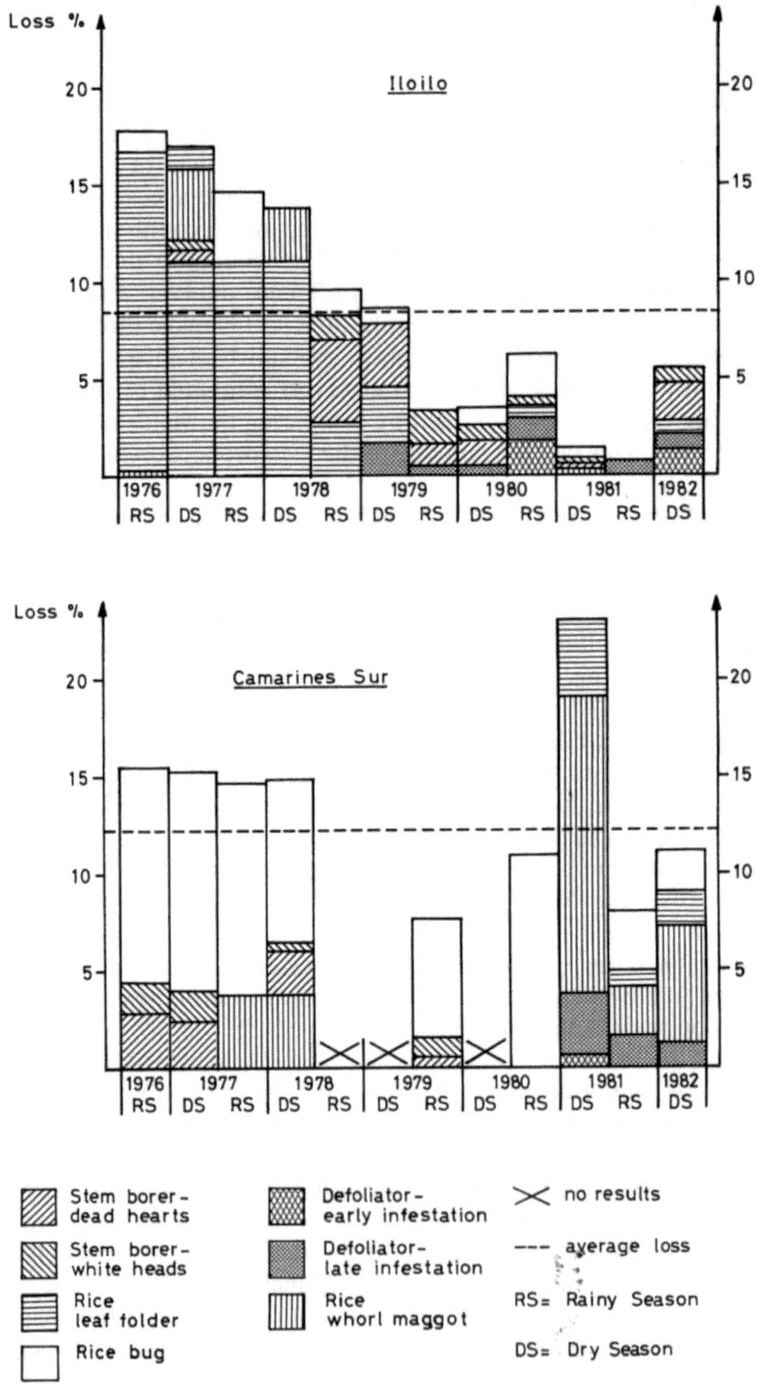

Fig. 6. Loss due to insect pests in resistant varieties in SEWS observation stations, two investigation areas, 1976–1982

Unfortunately, since no yield measurements were taken at the observation stations, we are unable to compare these crop loss figures derived from damage coefficients with the results of corresponding yield comparisons.

2.2.6.2.4 Summarizing Evaluation of Crop Loss Computations. The computation of crop losses on the basis of infestation data and damage coefficients leads to more accurate and especially more transparent results than a simple comparison of the yields of treated and untreated plots. Even yield differences, which by means of the customary tests are not found to be statistically significant, can be computed using this method. Furthermore, crop loss can be assigned to the pests responsible for it and presented graphically in a loss profile. Nevertheless, the test model used still gives rise to problems which can lead to an incorrect assessment of crop losses:

- The results depend to a high degree on the statistical quality of the functions constructed. The regression parameters as damage coefficients still have high confidence intervals (cf. Tables 17 and 19), and the computations based on them therefore possess a correspondingly high degree of uncertainty.
- Strictly speaking, damage coefficients should not be computed separately for individual pests, but calculated jointly for the main infestation combinations. Due to the limited data available, it was not possible to meet this requirement. On the other hand, if we add the loss computed on the basis of damage coefficients determined for single pests, an overestimation of crop loss may result.
- The methods used for measuring infestation levels were not identical in all cases, so that appropriate conversions had to be made. Such conversions perforce result in inaccuracies, particularly when infestation data expressed in scales of assessment are converted into percentages.

Although for the reasons mentioned above, the computations are still uncertain in some cases (rice bug), this does not apply to all the pests. The statistical test values obtained for the functions constructed are absolutely comparable to the corresponding figures found in the relevant literature. Many of the loss assessment functions computed in the United States have a coefficient of determination of only 0.2 to 0.7, and only rarely are better results achieved (MacKenzie and King 1980).

According to our above loss computations, crop loss in non-resistant varieties generally ranges from 12% to almost 40% depending on location. On the basis of these data, the tungro virus, responsible in one case for a loss of over 90%, must be regarded as possessing the highest loss potential. On the other hand, attention must be drawn to the fact that even in non-resistant varieties, losses only amount to less than 20% in some years.

In resistant varieties, crop loss is definitely lower, amounting to just over 10% on the average, with values of even less than 5% for some infestation events.

Overall, it is clear from the above results that the great difference in crop loss levels and in the composition of the damage caused make it necessary to prepare detailed economic analyses to identify the optimum strategy for a given objective.

3 Bases of Micro-Economic Analysis of Chemical Crop Protection Measures

Within the scope of integrated pest management, crop protection measures can be divided into preventive and curative measures (Palti 1981). Preventive measures mainly encompass the use of resistant varieties and appropriate cultural practices, and can also be described as measures aimed at reducing the probability of crop loss (Heitefuß 1975). Curative measures, on the other hand, comprise all direct pest control measures. From a decision-oriented point of view, crop protection measures can either be implemented prophylactically according to the generally expected damage levels, or in the form of targeted measures based on pest population monitoring.

In the case of rice cultivation, preventive measures include choosing the correct planting date, as well as selecting a variety suitable for the site concerned. However, these measures affecting crop protection will not be further discussed in the present study, since such decisions are greatly influenced by factors outside the pest complex. Furthermore, the possibility of changing these parameters is very limited in rice cultivation (Bernstein 1981). Planting dates, for instance, depend to a great extent on the availability of irrigation water and labour during the period concerned.

The economic analysis in the present study will therefore be limited to direct control measures. The problem involved in the economic analysis of these crop protection measures consists, in the final analysis, in determining the optimum control strategy. In the case of chemical measures, this is a question of determining:

- the optimum pesticide and method of application
- the optimum rate of application of the pesticide selected
- the optimum timing and frequency of application.

If the necessary information is available, the method of application most favourable from an economic point of view and the optimum type of pesticide/combination of pesticides can be determined by means of a simple comparison of costs and effectiveness. As regards the optimum rate of applications, there is only a very limited scope within the framework of the recommended minimum and the maximum rates fixed by law. Determining the optimum timing of a control measure, on the other hand, still involves considerable methodological problems. There is, in fact, an interaction between the minimum rate of application necessary to achieve a positive control effect and the optimum timing of a control measure. This interaction has led to the concept of the *economic damage threshold*. If the profitability of a crop protection measure is to be determined, the economic damage threshold must always be computed first since from an economic point of view, the latter is actually a break-even threshold. From the economic damage threshold, the *control threshold* can then be derived. Within the framework of existing control strategies, the above procedures represent possible alternatives with economic values which must be compared with those of the other possible methods.

3.1 Definition of the Concept of Economic Threshold

The concepts of *economic injury level* and economic threshold go back to Stern et al. (1959), the economic injury level being defined as the pest population density at which an increase would produce economic harm.

Economic threshold, on the other hand, was the term used for the pest population density at which a control measure should be implemented in order to prevent the economic injury level being reached. The difference between the two figures virtually consists only in a safety margin which has to be allowed to cover possible time lags between the time of infestation measurement and the time at which the control measure takes effect. Stern's et al. definition (1959) implies that both the economic threshold as well as the economic injury level refer to the point in time at which effective control is still possible.

However, it is often impossible at this early stage of plant growth studies to establish a relationship between the latter and the associated crop loss (cf. Chap. 2.2). It was the realization of this fact that caused Sylvén (1968) to introduce the concepts of "control threshold" and "critical injury level" to express in two stages the relationship between the target pest and the associated crop loss. A relationship is established between pest density, damage caused and crop loss. This relationship corresponds to a pest density which – in the case of non-control – would result at the end of plant growth in a crop loss which would justify control. The economic injury level, on the other hand, is based on the relationship between damage already existing, expressed in terms of infestation and crop loss. It can therefore only be employed for ex-post analysis of the economic feasibility of crop protection measures (Wüsten et al. 1981). The economic injury level is generally easier to determine than the control threshold for which data on pest population dynamics are necessary. The latter can consequently be described as a forecasting rule and is therefore always based on risk considerations.

3.1.1 Entomologists' Definition

The definitions given by Stern and Sylven do not deal in detail with the influence of economic factors. However, from their definitions it is quite clear that according to economic terminology, we are dealing with the break-even point, i.e. the yield level at which, according to the neoclassic production theory, the costs of the inputs required to achieve this yield are covered (Brandes and Woermann 1969). Assuming we have a technically fixed minimum application rate for a pesticide, certain fixed costs will result, which in turn will require the control measures employed to produce certain minimum returns. Using the damage coefficients, these returns can be expressed in terms of infestation units. This concept was defined more clearly by Norton (1976) and expressed by means of a simple formula:

$$\theta = \frac{c}{pdk},$$

where: θ = infestation intensity at which control is worthwhile
d = infestation intensity/crop loss relationship, i.e. the damage coefficient enabling the infestation to be assigned to a subsequent crop loss
p = price of product
k = effectiveness of control, expressed in terms of the proportional reduction in infestation intensity
c = control costs.

The expression *pdk* represents the returns produced by the control measure: when the control threshold/economic damage threshold is reached, the returns of the control measure must equal its costs. The above expression can refer either to the control threshold or to the economic damage threshold, depending on the time of plant growth or population development respectively for which the parameter *d* is defined.

The control thresholds and economic damage thresholds defined from the point of view of crop protection, and influenced by economic factors, indicate the specific infestation intensity at which a control measure is expedient. However, the level to which pest density should be reduced has hardly been discussed (Norgaard 1976). It has simply been assumed that the maximum kill rate is identical with the optimum one. This assumption was based on the fact that in the damage threshold concept, a given and accepted control measure was assumed. The input variation which is possible in practice (either by means of a change in the rate of application or by repeated applications during one and the same production period) has resulted in a different definition of the control threshold. This definition refers to the optimum rate of application and consequently to the economically optimum kill rate.

3.1.2 Economists' Definition

The above definitions of the economic damage threshold refer to the population density at which a control measure just begins to be profitable. However, no information is provided regarding the actual profitability of this measure. The latter depends on the monetary loss sustained in the case of non-control and on the technical effectiveness and the costs of the treatment.

In order to study this relationship in detail, it is necessary to abandon the assumption of one single control measure. The simplest case would consist in vary-

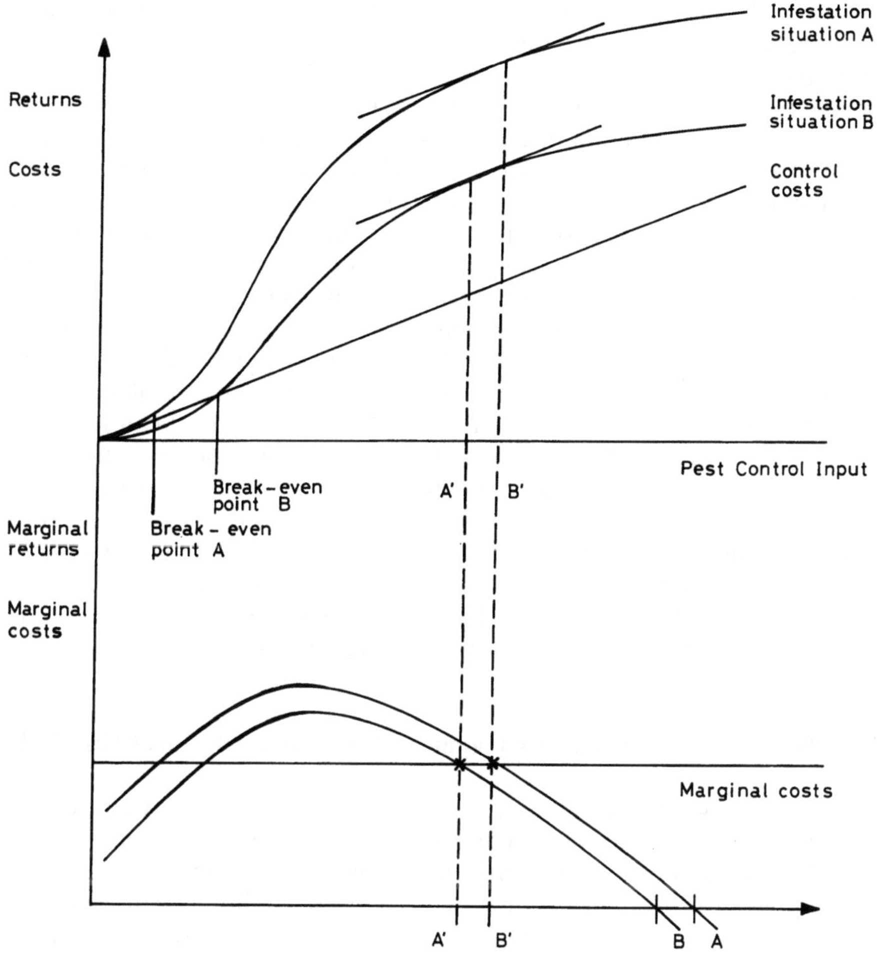

Fig. 7. Optimum level of pest control inputs

ing the rate of application of a certain pesticide. In this case, we can assume a changing relationship to exist between control costs and effectiveness of control. On the basis of a simple kill function of sigmoid shape, a positive effect, increasing at a decreasing rate until it reaches its peak, would be observed from a certain rate of application onwards (Bullen 1970).

The monetary loss prevented can be computed by relating the damage coefficient to the kill rate. Once this information has been obtained, the economically optimum dosage can be computed according to the principle of marginal productivity. This dosage will be reached when the incremental yield, obtained by increasing the kill rate and expressed in monetary terms (marginal returns), is equal to the incremental costs involved (marginal costs). Expressing this value in infestation units, we obtain the economically optimum pest density. This pest density corresponds to a different definition of economic threshold (Headley 1972a) which, however, is not at all unrelated to Stern's definition (Stern et al.

1959). Whilst the entomologists' definition refers to the break-even threshold, the economists' definition (Southwood et al. 1973) denotes the maximum monetary returns on a crop protection measure (Fig. 7).

The actual connection between the two definitions now consists in the fact that in both cases the optimum rate of application of the most suitable pesticide must first of all be determined. Once a control measure has been thus defined, the incremental yield – expressed in terms of infestation units – required to meet the costs of the measure corresponds to the entomologists' definition of the economic damage threshold. As long as there is only one practicable control alternative, this relationship can be disregarded, but in practice this would probably be the exception. If the recommended dosage has been geared to the maximum kill rate, which is frequently the case, a control threshold based on this figure would not necessarily lead to economically optimum results.

Since there is an economic damage threshold for every type of control measure, the optimum strategy for the generally expected infestation situation must first of all be determined. The economic damage threshold then corresponds to the break-even threshold of the optimum crop protection strategy. Furthermore, since the optimum application rate of a pesticide is influenced by kill rate and infestation intensity, it is necessary to determine the optimum dosage for the most frequent infestation intensities (Hillebrandt 1960). The various economic damage thresholds then have to be computed accordingly.

3.1.3 Relation Between Economic Damage Threshold and Control Threshold

From the definition of Sylvén (1968), it follows that control measures have to be implemented before the damage actually occurs. This poses a problem in as much as at this stage it is difficult both to detect the damage and/or to predict the resulting crop losses. It can be assumed that losses are all the easier to assess the nearer the pest comes to its damaging stage (cf. Chap. 2.1). In the case of plant diseases, this correlation is even clearer, since in this case a treatment applied at the time of visible damage, i.e. at the time the economic damage level is reached, comes too late. We must therefore define the control threshold as the infestation intensity which makes it likely – assuming pest progress takes place along the expected lines – that the economic damage threshold will be reached (Kranz and Hau 1981).

For insect pests, this definition also describes the existing correlations more clearly than the definition originally used by Stern (Stern et al. 1959). According to Dyck (1978), the use of control thresholds for rice pests can be regarded as a kind of short-term forecasting, although the currently used thresholds do not correspond to this definition in the case of insect pests. As shown by the analysis of the crop loss caused by insect pests (Chap. 2), most of the control thresholds used for rice pests are based on damage symptoms and not on pest density. However, if damage symptoms are used as a basis, we must be able to assume that a considerable part of the loss has not already occurred and that a control measure based on damage symptoms can still result in incremental yields of sufficient magnitude.

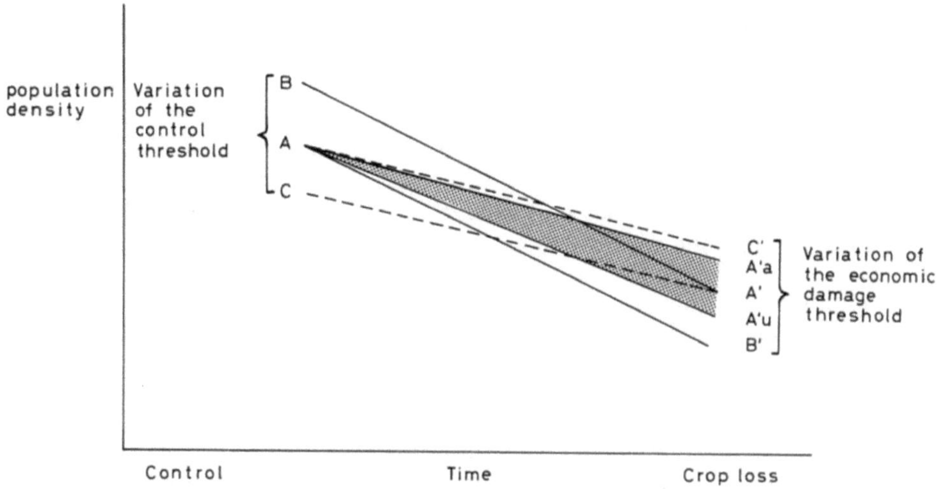

Fig. 8. Relationship between control threshold and economic damage threshold under different ecological conditions

From a statistical point of view, i.e. assuming that the damage can be prevented by one single application, the time at which infestation is measured is of utmost importance. The most appropriate time to measure infestation intensity and to determine the economic injury level is the time at which visible damage occurs. The decision on the control measure, on the other hand, must be taken earlier. Use of the control threshold concept generally requires infestation intensity to be measured at an early stage of pest development (or plant growth), although the pest does not enter its damaging stage until later, in accordance with its rate of multiplication. Accordingly, Chiang (1979) expanded the formula developed by Norton (1976) to include additional factors:

$$\text{Control threshold} = \frac{CC}{EC \cdot Y \cdot P \cdot YR \cdot SC} \cdot CF,$$

where:

CC = cost of control
EC = effectiveness of control
Y = potential yield
P = price of product
YR = damage coefficient
SC = survival coefficient, denoting the reproduction rate at the stage of damage occurrence of a post population measured at an early development stage (i.e. at the stage at which the population must be controlled). In Fig. 8, A denotes the population in its early stage of development and A' the same population at the time of crop loss. The slope of A–A' corresponds to the average rate of reproduction. A' also corresponds to the control threshold and A' to the economic damage threshold. Changes in

cultivation techniques will shift the relationship A–A' either towards A–B' or towards A–C'

CF = critical factor, permitting a risk-oriented or socio-economic valuation of the control thresholds. This is obtained by means of an ecologically founded variation of the relationship A–A' within $A'a$ and $A'u$. If, for instance, we were to select a CF value of 1 and an ecological situation unfavourable for pest development, the resulting crop loss would not suffice to meet the control costs.

In addition to the above factors, others may have to be taken into account. In situations in which the plant is under stress (e.g. lack of moisture), the damage coefficient and particularly the physiological damage threshold may be different than in normal situations (Chiang 1982).

The power of tolerance of a plant against a pest is variety-dependent. Furthermore, climatic factors may influence the SC factor mentioned above.

In addition, it must be taken into account that the relationship between infestation intensity at the various growth stages and the resulting crop loss can vary. In such a case, the quantitative relationship between the pest population at the time of control and at the time of occurrence of crop loss must be computed separately for all growth stages, i.e. in only a few cases will economic damage thresholds and control thresholds be parallel to the X-axis.

The information provided by Chiang (1982), which was verified by means of a number of examples, underscores the predictive character of control thresholds and thus basically confirms that the definition used for plant diseases by Kranz and Hau (1981) also applies to the field of entomology.

Since economic damage thresholds are multivariate functions with parameters which may assume different values from year to year and from field to field, they are valid only for a specific field and a specific year.

3.1.3.1 Relation Between the Two Threshold Levels for Rice Pests

As regards rice pests, it has already been emphasized that population measurements of most of the leaf pests and of the stem borer pose considerable practical problems. For this reason any control decisions based on damage must be related to the degree of infestation at the critical growth stage. Accordingly, an attempt can be made to relate the first measurements of damage symptoms to the infestation intensity at the critical stage. This will then permit the control thresholds to be derived from the respective economic damage thresholds. From a statistical point of view, we are thus using a two-stage method of assessment according to the above definition by Sylvén (1968). Using this method, infestation intensity at the critical development stage can be estimated in advance at the time of control by means of infestation criteria [Eq. (1)]. As endogenous variables, it would be possible to use the initial infestation intensity, the rate of growth between first and second infestation measurement, and climatological variables:

$$\overline{B}_K = f(B_A, r, K_1 \ldots K_n), \qquad (1)$$

where:

$\overline{B}_K$ = estimated infestation intensity at the time damage was caused
B_A = infestation intensity at the time of the first assessment
r = reproductive rate of the pest
$K_1 \ldots K_n$ = various climatological factors.

The above equation can only be solved by means of data from untreated plots, since only these data reflect the natural pest population development. The determination of the economic threshold is based on the relationship between infestation intensity at the critical stage and crop loss [Eq. (2)]:

$$EV = f(B_K), \qquad (2)$$

where:

EV = crop loss in %
B_K = measured infestation intensity at the critical stage.

Equation (2) will yield the factor d in the formula for computing the economic damage threshold. The latter will then serve as a measure of comparison for the result of Eq. (1). Thus, if the result of Eq. (1) is equal to the economic damage threshold the control threshold, is reached.

Equation (2) can be derived from field trials in which different degress of infestation intensity have been produced by means of chemical control measures.

It must be borne in mind that the test model presented in this study is justified only as long as there are no available damage coefficients representing several growth stages and enabling crop loss to be assigned directly to the causes responsible for it. This is because in a recursive method of estimation, as in the one being described here, the overall estimation error is always higher since the values which are substituted in Eq. (2) are already estimates in themselves.

3.1.4 Consideration of Mixed Infestations

In the case of infestation situations which are characterized by the occurrence of more than one pest, use of the threshold concept poses additional problems. When there is a positive correlation between the occurrence of several damage factors, the use of damage thresholds computed for individual pests may lead to economically wrong decisions. Furthermore, possible interaction between the different pests may result in the overall loss not being equal to the sum of the individual losses.

For this reason, damage coefficients must be computed by means of multiple regression equations, taking into account any possibly existing interactions:

$$Y = a - b_1x_1 - b_2x_2 \pm b_3x_1x_2,$$

where:

Y = actual yield
a = potential yield

b_1, b_2, b_3 = regression coefficients
x_1, x_2 = target pests
x_1x_2 = interaction between the two pests.

To determine the multiple threshold, the break-even yield must be determined in a second step:

$$\text{break-even yield} = \frac{\text{control costs}}{\text{overall control effectiveness in \%}} \times \frac{1}{\text{price of product}}.$$

Subtracting the break-even yield from the potential yield results in the above equation. We can then derive the minimum crop loss which would have to occur in the case of non-control when the economic damage threshold is reached. By solving the loss equation for x_1 or x_2, we obtain all the possible combinations of the two damage factors which will result in the economic damage threshold being reached.

To date, similar examples have not been presented in the relevant literature. In view of the complexity of the problems involved, this accepted principle so far has been to determine the economic damage threshold merely for the respective key pest (Stern 1973), although the question regarding the stimultaneous consideration of several pests is raised repeatedly (Talpaz and Friesbie 1975).

3.1.5 Dynamic Aspects

In the above discussion regarding the concepts of economic damage threshold and control threshold, the dynamic aspect of pest incidence has been neglected. However, if the decisions regarding the type of control measure to be implemented are to be realistic, the fact must be taken into account that there is an interdependence between control decisions taken during different production periods or at different times during one and the same production period. If by dynamic we understand the possibility of several control measures within one production period, then the quantitative effects of these treatment alternatives must be known. Assuming that the infestation intensity at the time of the treatment is shifted in time by one growth stage, then we can define the economic damage threshold for a specific stage as the infestation intensity at which the difference between the crop losses of the successive stages is equal to the break-even yield (Kranz and Hau 1981).

From a dynamic point of view, i.e. from the point of view of a number of production periods, basically, the same question is of relevance. A strategy which is regarded as optimum on its initiation, can become suboptimum on a long-term planning basis, e.g. if factors such as the increasing resistance of the pest to the pesticide used are taken into account (Hueth and Regev 1974). On the basis of a single-period approach, a control measure would be initiated at the point where the marginal returns cover the control costs involved.

The economic preconditions for a control measure will remain the same irrespective of whether a single or a multi-period approach is used. In the multi-period decision situation, we would consider the entire planning period and search out the strategy resulting in the maximum discounted overall returns, i.e. using a method similar to that proposed by Headley (1982).

3.1.6 Summary

The above comments regarding the concepts of economic damage threshold and control threshold are intended to indicate the complexity of the problems involved. However, not all the aspects dealt with in the theoretical discussion are relevant to the applied part of this study. This applies in particular to the dynamic aspects for which no data at all were available. The control thresholds still currently recommended by the official extension service have been taken into account as alternatives in determining the optimum control strategy. In addition, the control thresholds have been newly computed on the basis of the damage coefficients discussed in Chapter 2, taking into account the relevant economic factors. The effectiveness of control of the threshold values currently used can be derived from the empirical tests carried out to date. Whether the effectiveness of control of the newly computed thresholds is higher can again be determined only by means of trials conducted under field conditions.

A theoretically exact computation of control thresholds, which according to their economic definition must be equal to the optimum control strategy, is impossible at this point.

3.2 Methods of Calculation for Economic Analysis of Crop Protection Measures

In its widest sense, the economic analysis of crop protection measures aims at determining the optimum crop protection strategy for the infestation situation generally to be expected. Like analyses of other subjects of production economics, it involves the application of the economic criteria of production theory (e.g. break-even point and optimum level of input).

However, the analysis of the economic feasibility of crop protection measures possesses a higher degree of complexity. Thus, in crop protection analyses, the transparent presentation of the biological causes of the relation between input and crop yield is considerably more important than, for instance, in the case of fertilizer application. In contrast to fertilizer inputs, pesticide inputs only have an indirect effect on the yield, i.e. the control measure influences the pest complex and it is only the latter which in turn affects crop yields.

For this reason and due to technical interactions, the correct and economically optimum combination of the elements of a crop protection strategy, e.g. choice of pesticide, dosage, time and frequency of application, depends on the type and intensity of the pest infestation concerned.

Further influence factors at play in economic analyses of crop protection measures are the changing intensity and composition of pest infestation. At the time the farmer has to take a decision on the implementation of crop protection measures, he does not know what type of infestation event would occur in the case of non-control. He therefore has to make appropriate production decisions on the basis of incomplete information.

Due to the complexity of the input-output relationship, on the one hand, and the problem of uncertainty, on the other hand, relevant research studies are based on different concepts.

The main methodological possibilities will be outlined in the following.

3.2.1 Analysis of Production Functions

It is only under very special conditions that the assumption of a functional relationship existing between the implementation of crop protection measures and crop yields can be regarded as a suitable method for the economic evaluation of such measures. Provided we are dealing with only one damage factor (or with several pests which can be controlled in the same manner) and with an infestation varying only slightly in time, the optimum pesticide input level can be determined by employing the principle of economic margin. Computations of this kind were carried out by Hillebrandt (1960) for determining the optimum dosage for a given number of applications.

Other authors (Bartsch 1976; Ordish 1968) found a relationship between the frequency of application per cultivation season and crop yields. In both cases, the optimum rate of application, i.e. the rate at which marginal returns and marginal costs are equal, can be determined as a function of the respective price-cost ratio. As shown in Table 26, the relationship determined always applies to only one specific infestation intensity, so that the utility of such functions depends on the availability of data on the occurrence of the different infestation events. In the second case, the choice of the endogenous variables also poses a problem. In cases where a few applications suffice to produce maximum yields, the production function is difficult to determine. The effects of a change in the prices of inputs and outputs can only be estimated very approximately due to the fact that rounded figures are used.

If only one explanatory variable is employed, then the neoclassic production model can only be expected to yield useful results if composition and intensity of the damage complex are largely constant over time. To be realistic, several variables would always have to be used to define the returns to crop protection measures. However, these requirements cannot be met with the data currently available. Furthermore, it is virtually impossible to conduct the necessary multifactor trials with the simultaneous variation of all important factor combinations in the form customary to date.

One possibility of expressing the different inputs of a crop protection strategy by means of one variable consists in using monetary equivalents. In this method, solely the monetary expenditure on crop protection is employed as the explanatory variable of the output of crop protection measures. Attempts to determine the productivity of pest control were undertaken several times. Table 26 lists production functions in which the crop protection input is employed as explana-

Table 26. Regression coefficients of pest control variables in production functions for irrigated rice cultivation

Source	Unit	Regression coefficient	Total no. of variables	R^2
Herdt and Mandac 1979	Pesos $(\text{Pesos})^2$	− 1.688*** + 0.002***	32	0.658
	Pesos	+ 0.445***	16	0.698
Rosegrant 1977	Pesos	+ 1.47**	11	0.72
Marciano et al. 1981	No. of applications	+68.28$^{\text{n.s.}}$	6	0.34
Author's surveys in Nueva Ecija 1979	Pesos	+ 4.6*	4	0.45
Author's surveys in Nueva Ecija 1980	Pesos	+ 7.03**	3	0.52

*: Significant at $P \leq 0.05$
**: Significant at $P \leq 0.01$
***: Significant at $P \leq 0.001$
n.s.: Not significant

tory variable. Even if we disregard the statistical difficulties involved – in the first function the crop protection expenditure variable has a negative sign – it is clear that the explanatory value of such functions is small. Although the expenditure on crop protection has a significant effect on yield, we cannot automatically conclude that an increase in expenditure will have a positive effect on yield. On the contrary, it is always necessary to identify the elements of the crop protection strategy which are responsible for a certain level of expenditure. Thus, a mere change in the time of application may, for instance, increase the effect of a crop protection measure without any additional costs being involved.

However, the data necessary for incorporating several relevant explanatory variables are not available. Second, the limitation to simple input-output relationships would lend a rather pronounced "black-box" character to the production function analysis. The latter therefore does not appear to be a suitable analytical instrument for resolving the problems under consideration. The above analysis of infestation structure and loss assessment has thus shown that the preconditions for constructing production functions for pesticide input are not satisfied.

3.2.2 Cost and Returns Analysis

The disadvantage of production function analysis based on cross-sectional data can be overcome by setting up field trials. In such trials, different combinations (regarded as suitable on the basis of empirical data) of the above mentioned elements of crop protection strategies have to be tested as trial variants at the same location and continued for several years.

A first attempt in this direction was the trial series described under Chapter 2.1.3, to which the author has added trials of his own.

An economic analysis of such trials can be prepared by simply computing the difference between the costs and returns of the trial variants. Since these are clearly identifiable, the net returns (NR) can be computed as follows:

$$NR = (E_b - E_U)P - MK - AK - ZA - EK,$$

where:

E_b = yield after control
E_U = yield of untreated plot
P = price of crop at time of harvest
M_K = pesticide costs
AK = labour costs
ZA = cost of credit
EK = additional harvest costs due to higher yields.

A relevant example has been computed in Table 27, using data from pilot projects in Nueva Ecija, 1980, rainy season. In this example, the optimum control strategy would be method No. 4. The problems involved in the economic analysis must, of course, be viewed in close connection with the results of the statistical analysis of the trial. In the example selected, no significant yield differences could be found between treated and control plots. The yields listed must therefore be regarded as accidental since the statistical significance of the differences is not high enough to attribute them to the different control alternatives. In this case the logical conclusion to be drawn would be that non-control represents the optimum strategy. Although this fact is often ignored in economic studies (Herdt and Jajasuriya 1981), it is questionable whether such a mode of procedure is admissable. Furthermore, it has already been pointed out in the preceding chapters that the yield differences between alternative control measures cannot in every case be attributed to the differences in infestation intensity.

The objectives of the present study cannot be achieved by means of a simple economic analysis of the available trial results. Although these trials are basically suitable, they have one basic drawback, namely that the trial variants were not kept constant. The trials carried out by the author, on the other hand, do not suffice for a useful economic analysis.

Table 27. Costs and returns of alternative control strategies in Pesos ha^{-1}

Control strategy[a] (1)	Yield ($kg\ ha^{-1}$) (2)	Gross return (3)	Cost of pesticides (4)	Cost of labor (5)	Cost of credit[a] (6)	Cost of additional harvest[b] (7)	Total cost of control (8) = 4+5+6+7	Gross return minus cost of control (9) = (3)−(8)	Rank
1	4174	5008	194.5	21.0	15.1	59.4	290.0	4718.0	2
2	3645	4374	194.5	14.0	14.6	1.2	224.3	4149.7	6
3	3788	4545.6	119.0	14.0	9.3	16.9	159.2	4386.4	4
4	4397	5276.4	178.5	21.0	14.0	83.9	297.4	4979.0	1
5	3996	4795.2	178.5	21.0	14.0	39.8	253.3	4541.9	3
Complete protection	4436	5223.2	950.0	160.0	77.7	105.9	1293.6	4029.6	7
Untreated	3634	4360	0	0	0	0	0	4360.0	5
F-value	$1.6^{n.s.}$	–	–	–	–	–	–	–	–

[a] Computed from (5) + (6) on the basis of an interest rate of 7% for a term of 3 months (composite interest rate calculated from bank and private rates)
[b] According to current practice, harvest workers in the investigation areas receive about ⅙th of the crop yields as pay in kind
Source: Author's trials, Nueva Ecija, 1980 rainy season

3.2.3 Systems Analysis Approach

The above discussion regarding production function analysis and the economic evaluation of field trials by means of a computation of differences has demonstrated the limited analytical quality of these methods.

Since pest control contains biological as well as crop production elements, these have to be taken into account in the economic analyses.

Depending on the data available, either a detailed mathematical formulation is possible or a relatively simple assembly of the parameters of the main structural elements of the pest control system. The most accurate presentation of biological relationships is achieved in bio-economic simulation models, whereas in decision theoretic models we have to limit ourselves to the most important system parameters.

Irrespective of whether or not a detailed model is used, the following data always have to be known:

- yield level without the influence of the target pests
- losses caused by these pests
- availability and degree of effectiveness of alternative measures suitable from a technical point of view for controlling the relevant pests
- price and cost parameters
- objectives of the decision maker(s)

3.2.3.1 Bio-Economic Simulation Models

The obvious disadvantages of the above-discussed methods for the economic evaluation and optimization of crop protection strategies have resulted in the use of complicated procedures. The methods concerned are mathematical models for the quantitative presentation of biological and economic relationships. The aim of bio-economic simulation models always consists in maximizing the net returns or net benefit function. This function is connected via crop production aspects with the biological components of the system. An example of a model of this type (Reichelderfer and Bender 1979) is the Mexican bean beetle [*Epilachna varivestis* (Mulsant)]. The model concerned consists of six submodels:

- the population dynamics of the pest
- the growth of the crop plant
- the natural losses of leaf material
- the leaf material losses caused by the pest
- an infestation intensity/crop loss model
- the interaction between the target pest and its natural predator, *Pediobius foveolatus* (Crawford).

In the economic part of this model, the cost-benefit ratios of different biological and chemical measures were computed. Overall, 1479 different strategy-environment interactions were studied and biological control was found to be superior.

Although numerous models of this kind have meanwhile been constructed, they are always limited to either one single pest or to a small number of pests. If such models were to cover a larger number of pests, it would not only mean that the manifold possibilities of interaction could not be taken into account due to a lack of significant assumptions, but also that the limits of mathematical transparency would be reached. The analysis of the infestation situations and of the resulting crop losses has clearly shown that it is essential to take into account several pests occurring at the same time. The attempt to formulate an appropriate bio-economic model would consequently not be very promising.

3.2.3.2 Decision Theoretic Approach

Crop production decision situations can be classified according to the amount of information that is available on the assumptions on which the decision has to be based. If there is absolute certainty regarding the assumptions for the production decision, we speak of decisions under complete information (Brandes and Woermann 1969). On the other hand, if there is a discrepancy between the assumptions made at the time of the decision and the events actually occurring, we speak of uncertainty. This uncertainty can be broken down according to the degree of missing information (Weinschenck 1965). If the distribution of the events is known we are dealing with risk, if the probability of occurrence is unknown, we are dealing with a decision situation under uncertainty.

Crop production measures are a typical example of decisions taken under incomplete information. Since in most cases reliable forecasting methods do not exist, it is difficult to know which pest will occur and to anticipate their level of attack. For this reason, decision theoretic models are used to determine the optimum crop protection strategy.

According to Norton (1976), a decision problem in the field of crop protection economics is made up of the following elements:

- the possible infestation situations that would occur in the case of non-control (state of nature)
- the technical alternatives available for controlling the different infestation situations (control actions)
- the economic consequences of the different infestation intensity/control strategy combinations
- the decision rules used for evaluating the above combinations (utility).

The first two elements of the decision problem taken together form the third element which is normally presented in the form of a pay-off matrix.

The selection rules depend on the subjective aims of the decision-maker and on the degree of incompleteness of the information available.

3.2.3.2.1 Pay-Off Matrix. As explained in Chapter 2, the infestation situations occurring in the case of non-control can be estimated by means of the untreated plots of field trials. The crop losses resulting from these infestation events represent the upper limit of the returns on any crop protection measure. How much of this "potential loss" can be prevented will depend on the effectiveness of control of the existing control strategies.

The economic results of all possible infestation event/control strategy combinations may consequently differ from case to case. Generally, the economic coefficients (e.g. net return) of the strategy/environment combinations are presented in the form of a pay-off matrix (Fig. 9).

Fig. 9. Pay-off matrix

	Nature	Infestation events		
Decisions of the decision-maker	t	a_1	a_2	a_n
	Nature	Economic coefficients		
Control strategies	b_1	b_1a_1	b_1a_2	b_1a_{n1}
	b_n	b_na_1	b_na_2	b_na_n

The decision-maker consequently has a choice of n number of actions which, from a technical point of view, can be applied in all environmental situations. If the decision-maker selects strategy b_1 and event a_1 occurs, the ensuing economic result will be b_1a_1. The decision problem therefore consists in selecting the optimum strategy out of the economic results of b_na_m different environment/strategy combinations, i.e. the strategy in which case the environment/strategy combination yields the best economic result out of all the b_na_m possible environment/strategy combinations.

3.2.3.2.2 Theoretic Decision Rules. The information available for selecting a suitable pest control strategy is generally extremely incomplete. In most cases, the probabilities to be assigned to specific infestation events are unknown. Even if infestation time series covering several years are available, they are generally too short to derive likelihood probabilities of an infestation situation. Furthermore, biological as well as crop cultivation conditions are subject to continuous change. Every few years, different varieties with different kinds of resistance are introduced and in the long term, the farmers' cultivation techniques either lead to an increase or decrease in the importance of the various pests. Therefore, if the time period on the basis of which the likelihood probabilities have to be determined, is assumed to have a historic structure, wrong conclusions may be arrived at in the subsequent analyses (Anderson et al. 1977). At best, probabilities can be determined by means of a subjective estimate set up by the respective decision-maker. This, however, was not carried out because of the difficulty in assessing infestation events due to the complexity of pest problems. Furthermore, the strong influence of the chemical industry may lead to an overestimation of farmers' losses due to pest, so that subjective assessments were believed to be unrealistic.

Decisions on crop protection measures are therefore often decisions under uncertainty, as in the case of the situation here being studied. This has certain effects on the decision criteria to be applied. A detailed presentation of all known decision criteria will be dispensed with at this point. Instead, the criteria already employed in the relevant studies will be briefly discussed.

Crop protection measures are often implemented according to the insurance principle. This principle can be said to be based on the maximin rule (Mumford 1981). According to this rule, the farmer would select the strategy which would yield the best results in the worst possible situation. This rule meets the requirements of smallholders in developing countries who want to be sure of achieving a certain minimum income and of preventing crop failure rather than accumulating income.

Other decision rules such as the Savage-Niehans rule in its original form and in a modified version (Weinschenck 1965) have also been applied to the field of crop protection economics (Wüsten et al. 1981). However, the sole minimization of maximum or average opportunity losses does not seem to suffice as a basis for the production decisions of smallholders in developing contries since it fails to take into account the possible occurrence of extremely unfavourable events.

3.2.3.2.3 Practical Decision Rules. The theoretical decision rules presented above are unlikely to be used in their pure form by farmers in developing countries. It can, on the contrary, be assumed that the farmers base their production decisions on simple rules of thumb formed by their individual assessment of the economic consequences likely to result from the particular strategy/environment combination concerned. In this context, the so called safety-first principle plays an important role. According to this rule, the farmers' strategy will be aimed at minimizing the probability of occurrence of catastrophic situations.

This rule basically amounts to the kind of approach which calls for the definition of the term of "maximum admissible loss", and which, from the beginning, excludes any strategy likely to result in the non-attainment of a fixed minimum income. For computing the minimum incomes of the small rice farms, Roumasset (1976) used the following formula:

$$MEK = \frac{(MKB + DK - VG)}{F},$$

where: MEK = absolute minimum income
MKB = minimum expenditure for current consumption
DK = urgently repayable loan debts
VG = salable, non-vital possessions
F = area in ha.

The above approach is certainly not of universal validity, especially since the absolute minimum income should not be measured only in terms of subsistence requirements. Due to institutional and socio-cultural background conditions, sizeable income losses and resulting deterioration in the standard of living are often viewed much more negatively than objective assessment would warrant. For this reason, it seems necessary to determine a range for the necessary minimum income.

On the basis of the anticipated prices, we can express the minimum necessary income in terms of minimum yield. The latter is given by the following formula:

$$EG_u = \frac{EA + PA + EV + KK - AE}{F}$$

$$EG_o = EG_u + \frac{PK}{F},$$

where: EG_u = lower limit of minimum yield
EG_o = upper limit of minimum yield
EA = share given to harvest workers
PA = rent
EV = farm household rice consumption
KK = loan in the form of kg of rice, repayable on a basis short term
AE = income from non-farm sources, converted into kg of rice
PK = expenditure for current consumption converted into kg of rice
F = area in ha.

It will be noted that in the above formula all monetary parameters are expressed in terms of units of produce and refer to 1 ha. The lower limit of the minimum yield covers the items which either have to be deducted from the yield due to the socio economic conditions, or have to be regarded as essential for the continuation of the farm. If we assume that the farmer aims to maintain his average income level, then a realistic upper limit would include the average overall farm household expenditure on current consumption.

Constraints like these, where acceptance of a strategy is prevented by the farmers' behaviour, also have to be taken into account with respect to the economic coefficients of the actual strategy itself. If a specific pest control strategy involves the use of cash funds exceeding the farmer's financial possibilities, he will dispense with its application, even if the strategy concerned would lead to favourable results.

The limited availability of cash funds also explains the frequent use in developing countries of pesticide dosages lower than those recommended by the official extension service (Farrington 1972). When cash is limited, expenditures on pesticides compete directly with expenditures on fertilizer, thus giving rise to opportunity costs. If we assume that the cost of the most expensive measure used by the farmers in a particular area represents the upper limit of the acceptable costs of a crop protection strategy, then this further restricts the choice of strategies.

For the farmer, in accordance with his individual assessment of the economic consequences of the different strategy/environment combinations, all the strategies which do not violate the above restrictions are acceptable. Selection of the optimum strategy, on the other hand, will depend on the decision criteria used. If the probabilities of the events were known, then the strategy yielding the maximum expected net return could be selected. However, since they are not known, we can apply the Laplace rule (Bühlmann et al. 1975). According to this rule, the strategy resulting in the maximum average returns without violating the set restrictions would be chosen.

This kind of decision behaviour, which is also regarded as realistic in more developed agricultures, can be presented more simply: a farmer would choose a strategy which is safe in ensuring a minimum yield at an acceptable cost of crop protection (Norton 1980).

A more in-depth examination of decision behaviour, i.e. the elaboration of individual utility functions, would possibly lead to a better assessment of the realizability of crop protection strategies. However, a detailed study of this kind was not possible for want of appropriate.

3.2.3.3 Summary and Comparison

The methodological approach described above cannot be shown to be definitely superior. The choice of method will depend on the nature of the problems involved and the information available.

The use of bio-economic simulation models is always advisable in cases where a number of alternative control strategies are available and depend on a great many biological and technical factors. Suitable simulation models make it possible to identify the trial variants which are most relevant to the trial objective and which should therefore be experimentally tested in field trials. The decision theoretic approach, on the other hand, is more suitable when technical relationships are largely known, but the problem of uncertainty plays the major role. This is the case when composition and incidence of the pest complex to be controlled are subject to great change through time.

Frequently, when the nature of the problem to be investigated suggests that one or the other of the methods would be preferable, it is appropriate to combine elements from several methods. On the one hand, the biological and technical complexity of pest control makes it necessary to include these elements in economic analysis. On the other hand, decisions in the field of crop protection are almost always decisions under incomplete information. Optimization of the crop protection input in accordance with the basic tenets of the neoclassic production theory would consequently not correspond to actual requirements. Ideally, technical relationships should therefore be presented by means of simulation models whilst the economic results should be assessed on the basis of the principles of decision theory, taking into account the uncertainty factor. In his study, Carlson (1970) comes closest to meeting these requirements by combining, on the basis of the Bayes criterion, the subjective probabilities of the decision makers with the objective information obtained by means of a peach disease prediction model.

In the majority of cases in which similar problems are encountered in the developing countries, the data available do not suffice for a detailed approach of this kind. As a result, the availability of suitable data largely determines the type of methodological approach used for the economic analysis of pest problems.

4 Alternative Crop Protection Strategies in Irrigated Rice Cultivation: Comparison and Evaluation by Means of a Computer Model

In the preceding chapter, identification of the optimum crop protection strategy was presented as a decision theoretic problem under incomplete information. Figure 10 contains the main structural parameters of a corresponding model. With the exception of the damage coefficients dealt with in Chapter 2.2.6, all the necessary assumptions for this model will be discussed below. The main assumptions for this model will be discussed below. The main assumptions regarding the infestation situation and the economic parameters will be listed once more together with the results of the computations.

4.1 Description of the Model

Due to a certain set of environmental constellations which cannot be defined more exactly, a number of possible infestation events characterized by a specific combination of damage factors of varying intensity may occur in a given field. These infestation events on the one hand, compare with a discrete number of control methods, which, on the other hand, will be referred to as strategies in the following. Basically, a strategy is characterized by two economic parameters: the costs involved in its application and the effectiveness of control expected to result from it. The latter can be expressed in a more or less complex form, but can also be referred to more simply as the proportional reduction in the intensity of infestation by one damage factor at a specific time (Satapanthy and Aujaneyulu 1982). From this reduced infestation intensity, it is possible – by relating this with its corresponding damage coefficient – to derive the returns from a control strategy expressed in terms of produce. In addition, the economic results of a strategy/environment combination are affected by exogenous influences which from a micro-economic point of view, cannot be influenced by the decision-maker.

Each known strategy-environment combination leads to an economic result which – considered solely from the point of view of the crop protection system – can be regarded as a return, whereby return is defined as output expressed in monetary terms minus crop protection costs. The decision problem now consists in selecting the best strategy out of all the given *nj* combinations. The strategy to be regarded as the "best strategy" would be that which, taking into account both economic and non-economic restrictions, will contribute most to the achievement of a given objective.

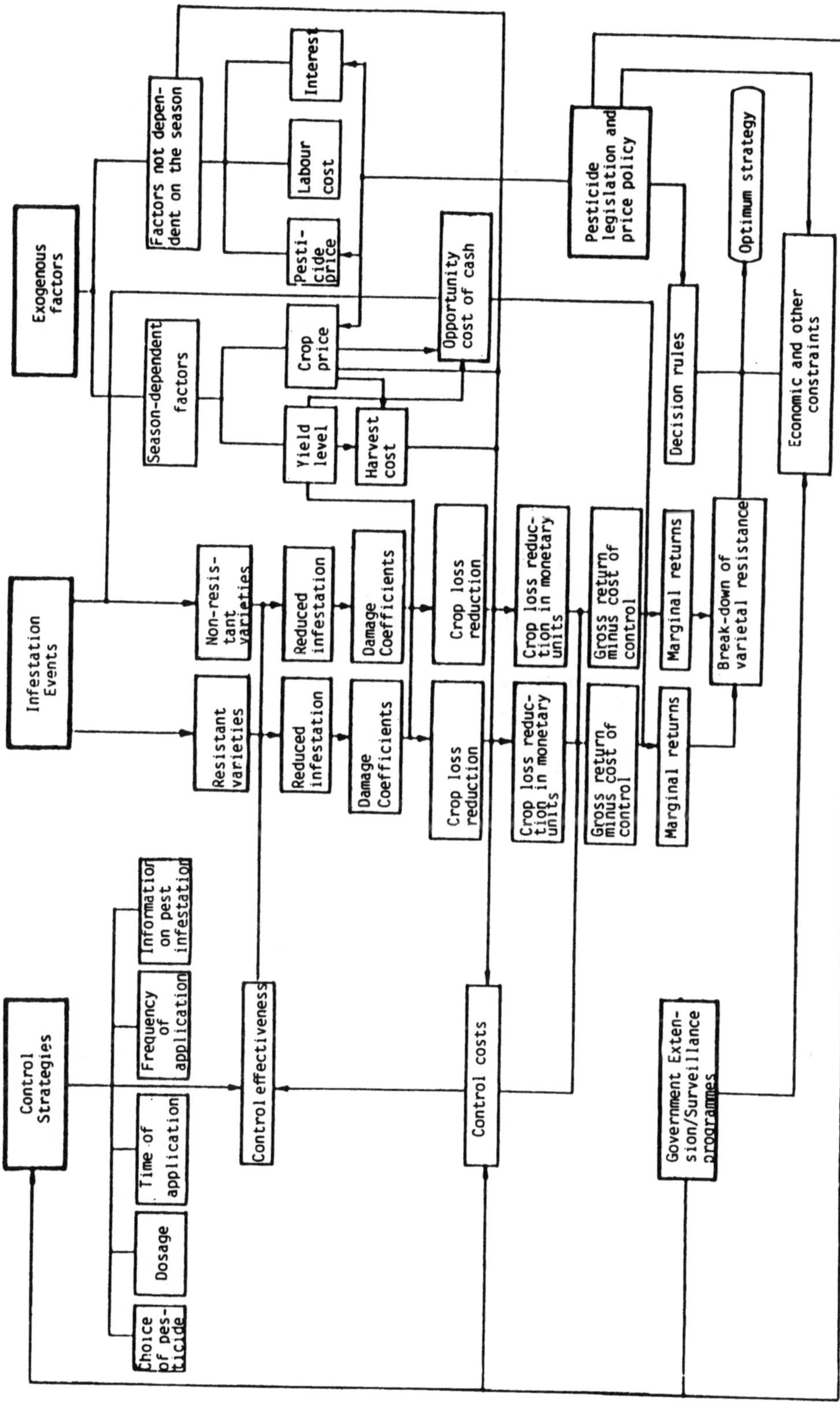

Fig. 10. Structure of a model for the determination of the optimum pest control strategy in rice

Outside of the micro-economic decision sphere, government policies can both affect the economic result of a strategy-environment situation and also limit/increase the number of strategies available for selection.

This type of model computation can be regarded as normative, since we are concerned mainly with the evaluation of alternative methods of chemical crop protection and not so much with the simulation of the actual control strategies used by the farmers. Nevertheless, in selecting the optimum strategy, certain constraints have been taken into account to ensure that actual field conditions are reflected as accurately as possible.

The results of the computer model provide information on the optimum control strategy for a given combination of infestation events and exogenous influences. From the results of the computations we can therefore derive extension recommendations for the individual investigation areas and also compute the differences between the benefits resulting from alternative crop protection strategies.

4.2. Model Assumptions

Chapter 2 gives the basic data necessary for computing damage coefficients. These data, obtained from a number of field trials, together with the loss functions computed from them, form the basis for part of the model assumptions.

Data on the effectiveness of the different pesticides and pesticide combinations for rice pest control were obtained from the pesticide evaluation trials. The parameters of the control thresholds recommended by the official extension service were derived from the tests carried out for this purpose. The data for describing the initial situation are provided by the trials simulating the farmers' main control strategies.

For the price and cost parameters, the official price statistics were consulted. In addition, actual farm-gate prices were determined by means of repeated surveys in the three investigation areas. These surveys also provided information on the liquidity situation of the farms in the investigation area.

From these surveys, it was also possible to derive the marginal yields of certain micro-economic production alternatives (e.g. different fertilizing and weed control methods). When there is a shortage of cash, these marginal yields represent the opportunity costs of pest control.

The present conditions of land tenure and the financial status of the farms form part of the economic constraints limiting the choice of control measures. Non-economic constraints, on the other hand, result from the subjective assessment on which the farmers base their choice of control strategy. These assumptions are incorporated in the model computations via the restriction of a minimum target yield. Furthermore, the results can be checked for their practicability by means of empirically determinable restrictions and plausibility considerations. This kind of information can also be derived from the surveys.

The individual assumptions will be explained in more detail below. As regards the presentation of infestation events and the computation of damage coefficients the reader is referred in Chapter 2.

4.2.1 Control Strategies

The present study is limited exclusively to chemical control measures for a detailed description of all the strategies used (see Appendix).

In accordance with the technical components of a crop protection strategy, e.g. choice of pesticide, dosage, timing and frequency of application, any combination of these components considered appropriate by technical standards can be defined as a control strategy. Basically, these strategies can be classified according to their most important component, i.e. the pesticide concerned. As shown by Fig. 11, the chemicals can be subdivided into granules, different sprays and combinations. Theoretically, the choice of pesticide which can be described as a discrete influence variable, can be combined with any number of variations of the continuous variables rate, time and frequency of application.

However, under field conditions, particularly in the developing countries, these variables are continuous only to a very limited degree. As regards choice of correct dosage, for instance, there is often only limited scope for variation, since the rate of application is fixed by the manufacturer and a lower rate might considerably reduce the killing effect. As regards fixing the time and frequency of application, biological factors or other farm activities (e.g. weeding) which have to be carried out at specific times often impose restrictions or give rise to opportunity costs.

In view of these difficulties, an exact mathematical presentation and construction of the pest control function are impossible. Instead, the methods which to date have been tested in trials will be employed as alternative strategies and from these results, specific pesticide combinations will be derived. In addition, the results of field trials simulating control strategies used by farmers in the different investigation areas can taken into account as alternatives.

4.2.1.1. Strategies for Granules

Due to their simple mode of application, granules have for many years formed part of the official extension recommendations. They can either be worked into the soil before transplanting or applied after transplanting together with the fertilizer. The extension service primarily recommends Carbofuran, not only because of its effectiveness, but also on account of its above-mentioned yield-stimulating effect.

Other important chemicals are Diazinon and Gamma BHC (Lindane). The latter is being used in particular against the stem borer. In a trial in which Carbofuran was applied after transplanting, eight combinations of timing, number of applications and dosage were tested. In the case of the two other groups of pesticide, several applications of only one input level were tested in insecticide trials. From these trials, three methods each can be derived: application directly after transplanting (early), application at the beginning of the generative phase (late) and a combination of the two. To simplify matters, it is assumed that the insecticide only affects the pests occurring during the particular growth stage during which it was applied. The possibility of a synergistic effect being produced in the case of combined application was neglected.

Overall, 14 different strategies for applying granules are obtained in this way.

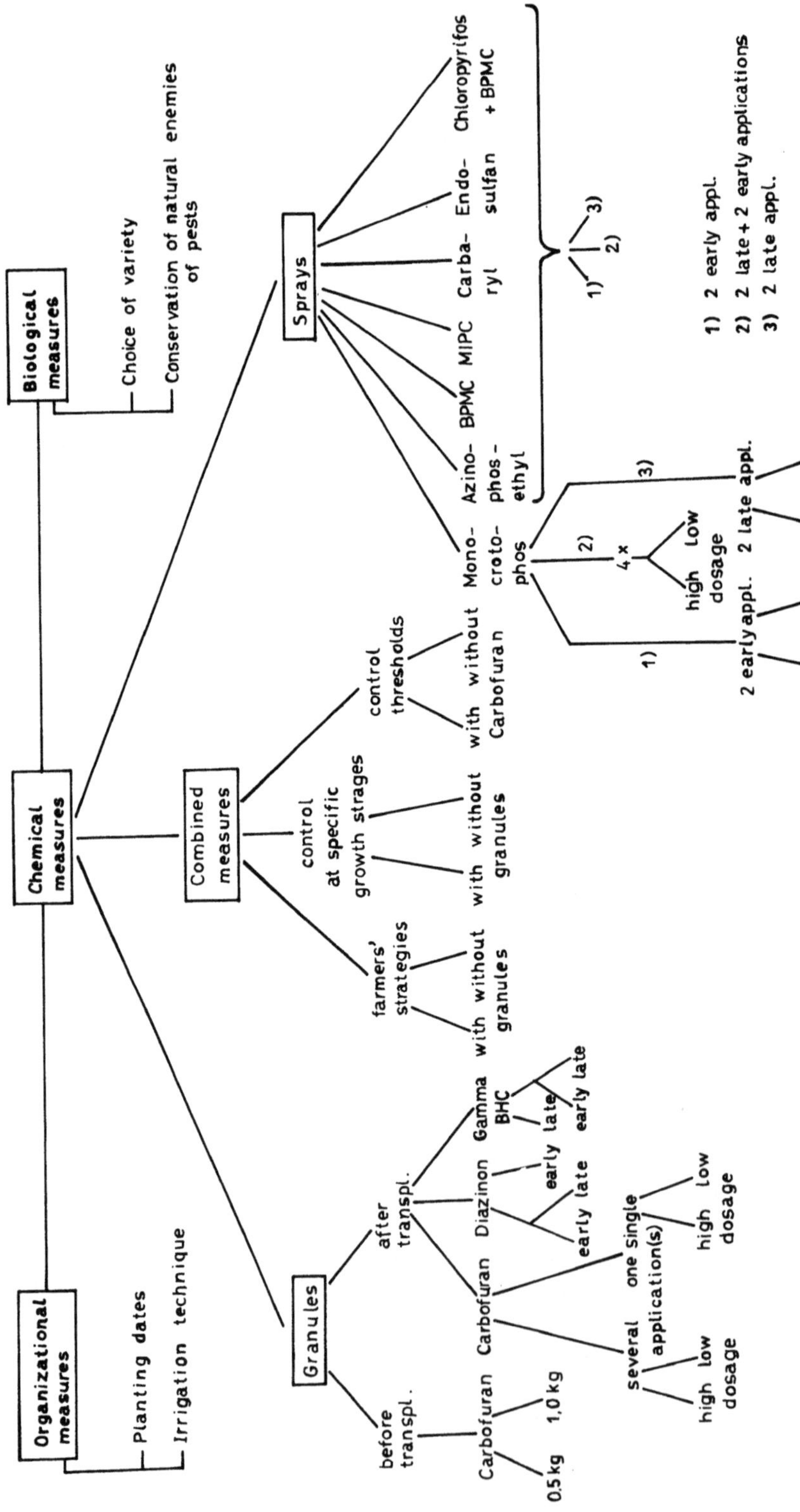

Fig. 11. Systematic presentation of crop protection measures

4.2.1.2 *Strategies for Sprays*

The seven groups of chemicals listed in Fig. 11 can be regarded as the insecticides most frequently used in rice cultivation. The only pesticide belonging to the group of the organochlorines is Endosulfan. The remaining pesticides belong either to the group of the organophosphates (Monocrotophos, Azinophosethyl) or to the carbamates (BPMC, MIPC, Carbaryl).

For the application of sprays, almost exclusively knapsack sprayers are used in rice cultivation. Although the official extension service recommends quantities of up to 1000 l ha^{-1} of spray mixture, the farmers – due to the high labour input involved – only apply 150 to 300 l ha^{-1} (Litsinger et al. 1978).

The particular method employed in the insecticide trials (i.e. four applications up to the generative growth phase) and the rates of application used result in three strategies for each group of chemicals: two early applications, two late applications and the four applications actually tested. In the case of Monocrotophos, a dosage of only 0.3 kg ha^{-1} was tested in addition to the usual dosage of 0.75 kg ha^{-1}. This resulted in a total number of 24 different spray strategies (Fig. 11).

Table 28. Main control strategies used by farmers in the three investigation areas

Strategy[a]	Pesticide	Dosage active ingr. (kg ha^{-1})	Time of applic. (DAT)	No. of applications
N1	Carbofuran Monocrotophos	0.5 0.17	5 25, 40	3
N2	Monocrotophos Carbofuran	0.17 0.5	10 30	2
N3	Monocrotophos	0.17	15, 25	2
N4	Monocrotophos	0.17	5, 15, 35	3
N5	Monocrotophos	0.17	10, 35, 60	3
C1	Azinophosethyl	0.2	15, 25	2
C2	Azinophosethyl	0.2	20, 40, 60	3
C3	Azinophosethyl	0.2	20, 40, 60, 80	4
C4	Azinophosethyl MIPC	0.2 0.5	15, 30, 45 80	4
I1	Chloropyrifos + BPMC	0.2	25, 45	2
I2	Chloropyrifos + BPMC	0.2	15, 35, 55	3
I3	Azinophosethyl	0.3	25, 45	2
I4	Azinophosethyl	0.3	15, 35, 55	3
I5	Carbofuran Chloropyrifos + BPMC	0.5 0.2	5 50	2

[a] N = Nueva Ecija, C = Camarines Sur, I = Iloilo
Source: Author's surveys

The assumptions to be made regarding the effects of strategies not tested in trials were derived in the same way as in the case of the granules.

4.2.1.3 Strategies Used by Farmers

The main difference between the strategies used in the field and the extension recommendations consists in the lower dosages employed in the former. According to IRRI investigations (Litsinger et al. 1978) in three Philippine Provinces, the customary rate of application is 0.03 to 0.38 kg a.i. ha^{-1}.

On the basis of an application rate of 0.75 kg a.i. ha^{-1} as recommended for most of the insecticides, this corresponds to an underdosing of 96 to 49%. The author's surveys in the three investigation areas yielded similar results as shown by the data in Table 28. However, the control strategies listed are not necessarily identical with the control strategies most frequently used by the farmers. Due to

Table 29. Control strategies recommended by the official extension service and by research agencies[a]

Growth phase	Pest	Pesticide	Dosage active ingr. (kg ha^{-1})	Time of application (DAT)	Type of recommendation
Vegetative	Rice whorl maggot	Carbofuran	0.5	Before transpl.	Prophylactic
Vegetative	Rice whorl maggot	Carbofuran or Diazinon	0.5, 1.0	5	Prophylactic
Vegetative	Rice whorl maggot	Monocrotophos, Azinophosethyl	0.75, 0.75	5 and 10	Prophylactic
Vegetative	Brown planthopper	BPMC, MIPC etc.	0.75	20 per hill	Control threshold
Vegetative	Green leafhopper (tungro)	BPMC, MIPC etc.	0.75	5 in 10 net sweeps if virus source exists	Control threshold
Vegetative	Rice caseworm and army worm	Carbaryl etc.	1.5	15% infested leaves	Control threshold
Vegetative	Stem borer	Chloropyrifos + BPMC etc.	0.75	15% dead hearts	Control threshold
Generative	Stem borer	Chloropyrifos + BPMC etc.	0.75	35 and 45	Prophylactic
Generative	Stem borer	Carbofuran	0.5	40	Prophylactic
Generative	Rice leaf folder	BPMC etc.	0.75	15% infested leaves	Control threshold
Ripening	Rice bug	Carbaryl etc.	0.75	4–6 m^{-2}	Control threshold

[a] In the case of control thresholds, the pesticides indicated are those used in field trials. The extension service does not recommend any specific products

Source: Interagency Pest Control Recommendation 1981 and Guidelines for a Surveillance and Early Warning System, unpublished papers, Manila 1981

the numerous possibilities of combining the elements of a pest control strategy, it is not always possible to identify the farmers' strategies exactly. Furthermore, control strategies change in the course of time, so that new strategies would have to be tested continuously.

For this reason, the strategies discussed here and tested in trials represent only the principal features (e.g. choice of pesticides, rate and time of application) of the control strategies commonly used in the field. Thus, for all three investigation areas, the most important pesticides at the above-discussed application rates were combined with different application dates within the chief pest control period (vegetative growth phase). In this way, five strategies are obtained for Nueva Ecija, four for Camarines Sur and five for Iloilo. In 3 out of the 14 strategies, Carbofuran is used.

4.2.1.4 Extension Recommendations

The measures recommended by the official extension service, formerly purely prophylactic, are now of a more curative type. Although the recommend measures are based on control thresholds, the application of Carbofuran before transplanting is recommended for most of the production sites. In areas with a high stem borer incidence, prophylactic spraying is also recommended during the generative growth phase. Table 29 summarizes the official pest control recommendations in rice dating from 1981 (unpublished report 1981). The measures concerned cover the main pests occurring during the various growth stages. For control of these pests, no specific pesticides are indicated, but a listing is given of all the pesticides which have been proved to be effective against the particular pests concerned.

Overall, this results in three prophylactic strategies and – corresponding to the number of existing control thresholds – seven further strategies, provided that only one threshold was reached each time. The control thresholds can again be combined with the use of Carbofuran prior to transplanting, so that a total of 17 strategies are obtained.

4.2.1.5 Other Strategies

The last group of strategies is based on plant growth stages and involves different levels of crop protection intensity. Table 30 lists the elements of these strategies. Since the pesticide input level is generally high in these strategies, they cannot be expected to be economically feasible. They are therefore purely hypothetical and should be employed only when the focus is on the production objective, i.e. when crop yields instead of net returns are to be maximized.

4.2.2 Effectiveness of Control

The data available permitted only a simple method of measuring the control effectiveness to be employed. The method concerned consists of a quotient (kill coefficient) which expresses the proportionate reduction in infestation intensity

Table 30. Control strategies based on growth stages

Type of strategy	Insecticides (common name)	Dosage (kg of a.i. ha^{-1})	Time of application (DAT)	No. of applications[a]
Complete pest control measure	Monocrotophos	0.75	5, 20 DAS	(2)
	Monocrotophos	0.75	5, 15, 25	
	Chloropyriphos + BPMC	0.75	40, 50	7
	Carbaryl	1.5	65, 75	
No control during ripening phase	Monocrotophos	0.75	5, 20 DAS	(2)
	Monocrotophos	0.75	5, 15, 25	
	Chloropyriphos	0.75	40, 50	5
No control during generative phase	Monocrotophos	0.75	5, 20 DAS	(2)
	Monocrotophos	0.75	5, 15, 25	
	Carbaryl	1.5	65, 75	5
No control during vegetative phase	Monocrotophos	0.75	5, 20 DAS	(2)
	Chloropyrifos + BPMC	0.75	40, 50	
	Carbaryl	1.5	65, 75	4

[a] Figure in parenthesis refers to applications in the seed bed
Source: 1980 IRRI Insecticide Report, Table 84 and author's computations

Table 31. Effectiveness of insecticides against different pests

Pesticides	Pests							
	GLH	BPH	SB	WM	LF	CW	AC	RB
Carbofuran before transpl.[a]	Yes	Yes	Yes	Yes	Yes	Yes	No	No
Carbofuran after transpl.[a]	Yes	Yes	Yes	Yes	No	No	No	No
Diazinon[a]	Yes	Yes	Yes	Yes	No	No[b]	No	No
Gamma BHC	No[c]	No	Yes	No	No	No	No	No
Azinophosethyl	Yes	No	Yes	No	Yes	Yes	Yes	No
BPMC	Yes	Yes	No[c]	No[c]	Yes	Yes	No	Yes
Carbaryl	Yes	No	No[c]	No[c]	No	Yes	Yes	Yes
Chlorophyrifos + BPMC	Yes	Yes	Yes	No	No	Yes	Yes	No
Endosulfan	Yes	No	Yes	No	Yes	Yes	No	Yes
MIPC	Yes	Yes	No[c]	No	No	Yes	No	No
Monocrotophos[a]	Yes	Yes	Yes	Yes	Yes	Yes[d]	Yes	Yes

[a] If used more than 40 days after transplanting, these pesticides may cause resurgence of the brown planthopper
[b] Is not listed to be effective, no trial results available
[c] Is listed as ineffective, but trial results nevertheless indicate a certain degree of effectiveness
[d] Is not listed to be effective, but trial results indicate a certain degree of effectiveness
Source: Interagency Pest Control Recommendations 1981

Table 32. Control effectiveness of strategies for granules[a]

Pest	Strategies													
	Carbofuran								Diazinon			Lindane		
	Before transplanting, one application after transplanting					After transplanting			Early	Late	Early + late	Early	Late	Early + late
	0.5 kg ha^{-1}	1.0 kg ha^{-1}	0.5 kg ha^{-1}	1.0 kg ha^{-1}	2.0 kg ha^{-1}	0.5 kg ha^{-1} (4×)	1.0 kg ha^{-1} (2×)	1.0 kg ha^{-1} (3×)						
Rice whorl maggot	0.78	0.74	0.22	0.27	0.33	0.22	0.51	0.36	0.45	0	0.45	0.33	0	0.33
Defoliator (veg. phase)	0	0	0	0	0	0	0	0	0	0	0	0	0	0
Rice caseworm	0.68	0.68	0	0	0	0	0	0	0	0	0	0	0	0
Defoliator (gen. phase)	0	0	0	0	0	0	0	0	0	0	0	0	0	0
Rice leaf folder	0.43	0.43	0	0	0	0	0	0	0	0	0	0	0	0
Stem borer (gen. phase)	0.42	0.42	0.38	0.47	0.47	0.47	0.64	0.64	0.8	0	0.8	0.77	0	0.77
Brown planthopper	0.43	0.63	0.43	0.43	0.43	(0.43)	(0.43)	(0.74)	0.63	(0.63)	(0.63)	0	0	0
Tungro	0.68	0.75	0.53	0.56	0.93	0.61	0.62	0.62	0.65	0	0.65	0.33	0	0.33
Stem borer (ripening phase)	0.4	0.6	0	0	0	0.59	0.57	0.8	0	0.72	0.72	0	0.76	0.76
Rice bug	0	0	0	0	0	0	0	0	0	0	0	0	0	0

[a] $\frac{\text{Infestation level in untreated plot} - \text{infestation level in treated plot}}{\text{Infestation level in untreated plot}}$

Source: Computed from insecticide trials, integrated pest management trials and author's trials

at the time at which the measurement is taken, which is identical with the point of time used in crop loss relationships. By relating this quotient to the damage coefficient, we obtain the loss prevented by the control strategy concerned, i.e. the return from the strategy expressed in terms of natural produce.

In the following, the author will discuss the kill coefficients – computed mainly from insecticide trials – for all the control strategies described above. The necessary basic data are taken from the pesticide evaluation trials. Table 31 lists the pesticides used in the strategies described, indicating the chemicals that are likely to be successful against the different pests.

4.2.2.1 Strategies for Granules

The kill coefficients listed in Table 32 for the strategies in which only granulates were used are, for the greater part, based on the results of trials. For the computation of some of the coefficients, data from only a few trials (but from a minimum of two) were available, so that these figures can be regarded only as reference values.

When a pesticide was noted by the official recommendations as being ineffective and trial results proving the contrary did not exist, control effectiveness was denoted by zero (Table 32). However, in the case of some of the pests, the results of the trials indicate a low to medium level of effectiveness. Lindane used against tungro, for instance, was found to have a control effectiveness of 33%. In the different Carbofuran strategies for tungro control, a single application at a higher dosage was found to have a positive effect. Although this effect disappears if the pesticide is applied several times after transplanting, a positive control effect is still produced against the pests occurring during the second growth stage.

In the case of the brown planthopper, the figures given in brackets in Table 32 are the coefficients of strategies in which the pesticide was applied more than twice, i.e. later than 40 days after transplanting. Research of the effects of different pesticides (Chap. 2.2.7.1.2) has revealed that these pesticides may lead to resurgence of the brown planthopper, thus reducing control effectiveness to zero and possibly even making it negative. In the model computations, an effectiveness of zero has therefore been assumed for these pesticides.

4.2.2.2 Strategies for Sprays

Table 33 lists the kill coefficients of strategies in which a single pesticide is applied by means of area spraying. Kill coefficients for the brown planthopper are shown to be lower than those for tungro control. This is in agreement with the information found in the relevant literature (Heinrichs 1978), according to which the effectiveness of area spraying for control of the brown planthopper ranges from 40 to 80%, although the effectiveness of most of the pesticides tested did not exceed 60%. For some pesticides, e.g. Monocrotophos, the results of the IRRI trials are largely consistent with the results of the official pesticide evaluation trials, 0.76 – IRRI, and 0.77 – pesticide evaluation trials. In the case of virus control, the kill coefficients can also be described as good, al-

Table 33. Control effectiveness of the strategies for sprays[a]

Pest	Azinophos-ethyl	BPMC	Carboryl	Chlorop-BPMC	Endosulfan	MIPC	Monocrotophos	
							0.75	0.3
Vegetative phase only								
Rice whorl maggot	0.53	0.35	0.4	0.54	0.36	0.16	0.48	0.48
Defoliator (veg. phase)	0.62	0.0	0.62	0.62	0.0	0.0	0.62	0.62
Rice caseworm	0.6	0.6	0.6	0.6	0.0	0.6	0.6	0.6
Defoliator (gen. phase)	0.0	0.0	0.0	0.0	0.0	0.0	0.0	0.0
Rice leaf folder	0.0	0.0	0.0	0.0	0.0	0.0	0.0	0.0
Stem borer (gen. phase)	0.67	0.53	0.62	0.81	0.6	0.5	0.74	0.65
Brown planthopper	0.57	0.58	0.22	0.68	0.55	0.6	0.77	0.76
Tungro	0.91	0.72	0.59	0.91	0.6	0.36	0.84	0.83
Stem borer (ripening phase)	0.0	0.0	0.0	0.0	0.0	0.0	0.0	0.0
Rice bug	0.0	0.0	0.0	0.0	0.0	0.0	0.0	0.0
Generative phase only								
Rice whorl maggot	0.0	0.0	0.0	0.0	0.0	0.0	0.0	0.0
Defoliator (veg. phase)	0.0	0.0	0.0	0.0	0.0	0.0	0.0	0.0
Rice caseworm	0.0	0.0	0.0	0.0	0.0	0.0	0.0	0.0
Defoliator (gen. phase)	0.66	0.0	0.66	0.66	0.0	0.0	0.66	0.66
Rice leaf folder	0.81	0.85	0.0	0.0	0.0	0.0	0.81	0.81
Stem borer (gen. phase)	0.0	0.0	0.0	0.0	0.0	0.0	0.0	0.0
Brown planthopper	0.58	0.58	0.22	0.68	0.55	0.6	0.77	0.76
Tungro	0.0	0.0	0.0	0.0	0.0	0.0	0.0	0.0
Stem borer (ripening phase)	0.63	0.66	0.61	0.68	0.51	0.12	0.64	0.54
Rice bug	0.43	0.0	0.45	0.0	0.0	0.0	0.43	0.43
Vegetative and generative phases								
Rice whorl maggot	0.53	0.35	0.4	0.54	0.36	0.16	0.48	0.48
Defoliator (veg. phase)	0.62	0.0	0.62	0.62	0.0	0.0	0.62	0.62
Rice caseworm	0.6	0.6	0.6	0.6	0.0	0.6	0.6	0.6

Table 33. Continued

Pest	Azino-phos-ethyl	BPMC	Car-boryl	Chlor-op-BPMC	Endo-sulfan	MIPC	Monocroto-phos	
							0.75	0.3
Defoliator (gen. phase)	0.66	0.0	0.66	0.66	0.0	0.0	0.66	0.66
Rice leaf folder	0.81	0.85	0.0	0.0	0.0	0.0	0.81	0.81
Stem borer (gen. phase)	0.67	0.5	0.62	0.74	0.6	0.5	0.74	0.65
Brown planthopper	0.81	0.72	0.71	0.74	0.6	0.71	0.77	0.6
Tungro	0.91	0.72	0.59	0.91	0.6	0.16	0.64	0.86
Stem borer (ripening phase)	0.63	0.66	0.61	0.68	0.51	0.12	0.64	0.94
Rice bug	0.43	0.0	0.43	0.0	0.0	0.0	0.43	0.43

[a] Control effectiveness: $\frac{\text{infestation level in untreated plot} - \text{infestation level in treated plot}}{\text{infestation level in untreated plot}}$

Source: see Table 32

though it remains an open question as to whether it will be possible to achieve an effectiveness of control of over 90% even in the case of very high infestation.

As regards the rice whorl maggot, the level of control effectiveness arrived at by means of spraying is low throughout (16–54%). In the case of the remaining pests, the level of effectiveness attained using the pesticides listed as being effective (Table 33) ranges from 60 to 70%.

4.2.2.3 Strategies Used by Farmers

On testing farmers' strategies in field trials, their effectiveness was generally found to be lower than the kill coefficients arrived at in the insecticide trials (Table 34). For the stem borer (generative phase), for instance, the average effectiveness of all the control strategies used by the farmers was computed to be 39%, the corresponding figure from insecticide tests being 64% for sprays and 59% for granules.

The assumptions to be made regarding brown planthopper and tungro virus control posed a problem. Since only resistant varieties were used in the field trials (IR-36, IR-42, IR-54), the effectiveness of control in this case could not be tested.

In view of the fact that tungro epidemics and mass reproduction of the brown planthopper occurred despite the farmers' prophylactic measures, it can be concluded that the kill rate achieved during pest control was low. Only when the far-

Table 34. Effectiveness of control strategies used by farmers[a]

Pests	Strategies[a]													
	N1	N2	N3	N4	N5	C1	C2	C3	C4	I1	I2	I3	I4	I5
Rice whorl maggot (gen. phase)	0.22	0.54	0.12	0.35	0.36	0.26	0.32	0.32	0.32	0.28	0.35	0.39	0.29	0.15
Defoliator (veg. phase)	0.76	0.3	0.77	0.49	0.28	0.31	0.54	0.54	0.86	0.5	0.51	0.52	0.57	0.67
Rice	0.36	0.13	0.24	0.25	0.33	0.31	0.54	0.54	0.86	0.41	0.2	0.48	0.34	0.48
Defoliator (gen. phase)	0.41	0.11	0.32	0.25	0.17	0.39	0.82	0.82	0.81	0.37	0.49	0.68	0.58	0.53
Rice leaf folder	0.48	0.0	0.19	0.69	0.1	0.18	0.80	0.80	0.89	0.56	0.74	0.53	0.65	0.31
Stem borer (gen. phase)	0.26	0.27	0.18	0.3	0.44	0.39	0.64	0.64	0.54	0.47	0.30	0.35	0.22	0.41
Brown planthopper	0.0	0.0	0.0	0.0	0.0	0.0	0.0	0.0	0.0	0.0	0.0	0.0	0.0	0.0
Tungro	0.53	0.0	0.0	0.53	0.0	0.53	0.0	0.0	0.0	0.0	0.0	0.0	0.0	0.53
Stem borer (ripening phase)	0.18	0.17	0.11	0.18	0.24	0.25	0.25	0.78	0.83	0.24	0.34	0.34	0.37	0.42
Rice bug	0.01	0.22	0.11	0.18	0.28	0.0	0.19	0.33	0.56	0.0	0.24	0.0	0.24	0.24

[a] Control effectiveness: $\frac{\text{infestation intensity in untreated plots} - \text{infestation intensity in treated plots}}{\text{infestation intensity in untreated plots}}$

[b] For details of strategies see Table 28

Source: Author's trials in collaboration with the regional Crop Protection Centers 1980–81

mers applied Carbofuran shortly after transplanting could the coefficient given in Table 32 (53%) be assumed. The same applies to spraying a suitable pesticide twice after transplanting. However, in the case of the brown planthopper, due to the farmers' method of spraying (the lower part of the plant is virtually not reached), a very low effectiveness of control must be reckoned with. To simplify matters, a value equal to zero can therefore be assumed. Although this assumption results in the farmers' strategies being undervalued, even a low effectiveness of control (10–20%) would still lead to crop losses incompatible with the objective of farmers still largely operating on a subsistence basis. These strategies are consequently unacceptable irrespective of their benefit-cost ratios.

For all the strategies differing only by one additional application (e.g. C_2 and C_3), the mean kill coefficients for pest insects of vegetative and generative phases are listed in Table 34.

In the case of strategies I_2, I_4 and I_5, the trial results did not yield any coefficients for the rice bug. For this reason, the mean values of the other strategies were again used.

4.2.2.4 Extension Recommendations

As regards the strategies recommended by the extension service, the effectiveness of measures consisting of prophylactic calendar-based spraying can be derived from the results of the official pesticide evaluation trials. For extension recommendations based on control thresholds, on the other hand, it is necessary to fall back on a small number of trials in which a single pest reached its control threshold. The economic evaluation of a large number of IRRI trials reveals that control based on control thresholds often results in only minimal incremental yields which may not suffice to cover control costs (Herdt and Jajasuriya 1981). In Table 35, an economic analysis and yield comparisons have been prepared for the trials published in the IRRI Insecticide Reports, in which different pesticide input levels were used. Only results from trials in which one pest had reached its control threshold were included. Although incremental yields were obtained in three cases, these were still clearly lower than the potential yields. In two cases, control based on thresholds even resulted in a yield decline. Control effectiveness can be computed by relating the crop loss occurring in non-control to crop loss in control based on thresholds. The average potential loss in the five trials amounted to approximately 10%, with a residual loss of about 7%. This would mean that treatments based on control thresholds can be assumed to result in an effectiveness of control of 30%.

There is no doubt that the above figure can be regarded only as a rough indication, although it is basically consistent with the values given by the authors quoted and therefore appears to be realistic.

Table 35 also lists the cases in which two pest species reach their control thresholds. However, when applications take place at short intervals, these strategies again assume the nature of calendar sprayings and their effectiveness is greater. By relating the residual loss (5.67%) to the potential loss (22.88%) as before, we obtain an effectiveness of control of the order of 75%.

Table 35. Yields achieved (kg ha^{-1}) using alternative control strategies for pest situations in which control threshold is reached by one or two pests

Control threshold reached by: one pest						Control threshold reached by: two pests					
Potential yield[a]	Yield ETL[b]	Yield untreated	Potential loss[c]	Residual loss[d]	Source	Potential yield[a]	Yield ETL[b]	Yield untreated	Potential loss[c]	Residual loss[d]	Source
6005	5840	5720	5.70	3.47	IRRI Insect. Eval. Report 1980, Table 59	4020	3840	3770	6.2	4.47	IRRI Insect. Eval. Report 1980, Table 60
3820	3700	3710	2.88	3.14	Ditto, Table 68	4230	3980	3760	11.11	5.91	Ditto, Table 68
5210	4750	5120	1.73	8.83	Ditto, Table 69	3820	3820	2280	40.31	0.0	Ditto, Table 77
3040	2680	2160	28.95	11.85	Ditto, Table 70	7180	6390	4810	33.0	11.0	Insect. Eval. Report 1981, Table 61
4686	4294	4174	12.6	10.93	Own trials, Iloilo 1981/82	4692	4366	3575	23.81	6.95	Author's trials, Camarines Sur 1980
$\bar{x}$			10.37	7.64					22.88	5.67	

[a] Yields achieved at maximum pesticide input (maximum protection)

[b] Yields achieved by means of treatments on control thresholds

[c] $\frac{\text{Potential yield} - \text{yield of untreated plot}}{\text{potential yield}} \times 100$

[d] $\frac{\text{Potential yield} - \text{yield achieved by means of control thresholds}}{\text{potential yield}} \times 100$

Table 36. Control effectiveness of strategies based on growth stages[a]

	Complete treatment	No treatment during:		
		Ripening phase	Generative phase	Vegetative phase
Rice whorl maggot	0.6	0.6	0.6	0.0
Defoliator (veg. phase)	0.93	0.93	0.93	0.0
Rice leaf roller	0.64	0.64	0.64	0.0
Defoliator (gen. phase)	0.95	0.95	0.56	0.27
Rice leaf folder	0.93	0.93	0.72	0.87
Stem borer (gen. phase)	0.97	0.97	0.67	0.12
Brown planthopper	0.54	0.54	0.28	0.43
Tungro	0.96	0.96	0.95	0.0
Stem borer (ripening phase)	0.95	0.87	0.34	0.86
Rice bug	0.6	0.0	0.47	0.6

[a] For description of strategies refer to Table 30

Source: Author's trials in collaboration with the regional Crop Protection Centres, in Regions III, V and VI; author's calculations and IRRI Insecticide Evaluation Report 1979–1981

4.2.2.5 Other Strategies

With regards to effectiveness, the strategies aiming at the complete insecticidal protection of one or more growth stages must be regarded as superior to all other strategies (Table 36). These strategies will therefore be most effective in minimizing crop losses, the objective pursued by the decision maker. However, the costs involved are extremely high. We will nevertheless take these strategies into account as alternatives in our model calculations for obtaining data on their competitiveness in situations in which extremely high pest densities are encountered.

4.2.3 Control Costs

Basically, the costs of applying a crop protection strategy are made up of the costs of the pesticide used and of its application. Pesticide and labour costs (unless family labour is used) result in interest costs. In the case of borrowed capital, corresponding interest rates have to be taken into account; if personally owned capital is used, the interest rates for short-term savings deposits will have to be applied. Added to this, there are the costs of harvesting the incremental yield resulting from the crop protection measure.

Table 37. Prices of insecticides frequently used by farmers (in Pesos)

Pesticide	Extension service		Nueva Ecija		Camarines Sur		Iloilo	
	A	B	A	B	A	B	A	B
Granules								
Carbofuran	135.0	270.0	142.2	284.4	–	–	130.0	260.0
Diazinon	76.5	153.0	–	–	–	–	–	–
Lindane	48.0	96.0	75.0	150.0	–	–	–	–
Sprays								
Azinophosethyl	75.3	113.0	86.6	129.9	83.3	124.9	95.7	143.6
BPMC	56.0	84.0	–	–	–	–	68.5	102.8
BPMC + Chloropyrifos	116.0	348.0	112.8	338.4	119.8	357.0	115.6	346.8
Carbaryl	28.7	86.0	–	–	–	–	32.1	96.3
Endosulfan	66.7	100.0	63.3	94.9	74.7	112.0	69.5	104.2
MIPC	54.0	81.0	63.5	95.3	68.8	103.2	80.7	121.1
Monocrotophos	175.0	262.0	157.7	236.6	162.2	242.7	194.2	290.7

A: Based on 0.5 kg active ingredient
B: Based on the officially recommended dosage
Source: Interagency Pesticide Recommendations 1980 and author's survey

4.2.3.1 Pesticides

The various insecticides used in irrigated rice cultivation in the Philippines differ in unit price. Within each group of chemicals there are a number of different commercial products, the prices of which differ only slightly.

In our analysis, the price of the most frequently used pesticide within each group has been used (Table 37). On the average, the prices of the pesticides recommended within the framework of the National Rice Production Program are 10% lower (Magallona 1980). At the farm level, considerable price variations are sometimes encountered, since the dealers frequently grant supplier credits and add the interest to the price.

4.2.3.2 Labour

Sprayable pesticides are applied by means of hand-operated knapsack sprayers. Variable machinery costs do not arise, and water, the only input necessary for applying the pesticides, can be freely taken from the irrigation canals or the paddy fields. Human labour is consequently the only cost factor involved. Going by the farmers' customary rates of application, a labour input of 1 man-day ha^{-1} per application can be assumed. If hired labour is employed, a full man-day is normally counted when more than half a day's work is involved. For the rates of application recommended by the official extension service, about 2.5 man-days ha^{-1} per application have to be assumed.

The wage rate is clearly below the minimum wage fixed by the Government. According to the surveys of the Bureau of Agricultural Economics (Bureau of Agricultural Economics 1980), the wages actually paid were of the order of 6 Pesos per man-day as opposed to the official minimum wage of 15 Pesos. According to the author's own surveys in the three areas under consideration, 7 Pesos were paid on average. Normally, food is also provided for hired labour, so that the overall wage can be assumed to amount to 10 Pesos per man-day.

Since the farmers are in the habit of interchanging any free labour capacity available, the wages paid to outside labour can also be rated as opportunity costs of family labour.

4.2.3.3 Interest

Smallholdings with incomes mainly derived from rice production are extremely short of cash funds at the beginning of each planting season. The farms are therefore often forced to borrow money to buy inputs. Such loans may be obtained from institutionalized or non-institutionalized sources.

At the beginning of the national Masagana 99 rice production programme, Government-secured bank loans were very widespread. Today, due to the low rate of repayment, coupled with the administrative delays in the granting of loans, this type of loan has become a less important source of cash (Barlow et al. 1983). Loans from dealers, relatives or other private sources, on the other hand, are more frequent. In the past several years, the rural banks set up within the scope of the Masagana programme have also started to grant loans to smallholders, part of the working assets of the latter (e.g. water buffalo) being used as security.

The interest rates of the loans granted within the scope of the rice production program were subsidized and amounted to 12% p.a. Non-institutionalized bank loans are always based on the capital market rate of interest which in 1980 averaged 22%. However, the farmers often obtained different rates.

Loans from private individuals are based on agreements which can vary greatly. One customary form is the credit in kind consisting of two bags of rice at planting time to be repaid by three bags of rice after the harvest. For cash loans, interest rates of 10% per month are regarded as normal.

For loans from relatives, interest often does not have to be paid (Ledesma 1982) since there is a mutual obligation to help in emergencies.

To determine the rates of interest that actually have to be paid by the farmers, we have to compute the average rate and weight it according to the proportion of loans obtained from the different sources. To simplify this computation, the loans were differentiated only according to bank and non-bank loans.

From the socio-economic surveys carried out in the three investigation areas, i.e. in Nueva Ecija, Camarines Sur and Iloilo, the weighted rates of interest of 55.3%, 43.1% and 59.8% p.a. respectively were computed. Relating these figures to the average term of 3 months for which the loans are taken out, we obtain interest rates ranging from almost 11% to nearly 15% per cropping season, depending on the cultivation area concerned.

4.2.3.4 Cost of Additional Harvest

In the Philippines, rice is still harvested by hand. Traditionally, landless village inhabitants are employed for this operation (Hayami 1978). The harvest workers also carry out the threshing, for which small threshing machines are used. Workers are usually paid in relation to crop yields. In the areas investigated, the wages in kind paid to harvest workers were roughly equal to one ninth of the grain yield. The incremental yields achieved through crop protection measures consequently have to be reduced accordingly.

4.2.4 Socio-Economic Conditions in the Three Investigation Areas

In the introduction, a brief description was given, based on the national statistics, of the current position and the institutional framework conditions prevailing in irrigated rice farming. Since the current conditions affect the determination of the optimum crop protection strategy, they have to be taken into account for the model assumptions. For this reason, the conditions prevailing in the three investigation areas will now be briefly described. In this context, it is necessary to draw attention to the connection between the conditions of land tenure, farm sizes and the liquidity situation of the farms, the latter being of considerable importance in the selection of control strategies. When the liquidity situation of the farm is unfavourable, use of the working capital for crop protection will compete with its use for other operations or inputs, e.g. fertilizer, weed control or certified seed.

4.2.4.1 Farm and Family Size

From cropping acreage and family size, the land-man ratio can be derived, which can be regarded both as an expression of land productivity and also as a measure of population pressure. Given a certain yield level and non-farm income, the land-man ratio then provides information on the income situation of the farm. The latter in turn will determine the magnitude of the financial resources available for years with extremely low yields – such as may result from pest calamities – and therefore also the willingness of the farmers to take risks. Furthermore, it also determines the cash funds available to the farms for financing crop protection strategies involving a high cash input.

Table 38 gives the coefficients of farm and family sizes for the three investigation areas. With respect to acreage per farm, relatively clear differences between the three investigation areas can be seen.

Table 38. Coefficients of farm and family size in the three investigation areas

Area (province)	Farm size				Family size			
	$\bar{x}$	s	min.	max.	$\bar{x}$	s	min.	max.
Nueva Ecija	2.29	0.76	0.5	4.0	6.9	2.8	2	15
Camarines Sur	1.62	1.05	0.5	5.0	7.3	3.05	2	15
Iloilo	1.34	0.71	0.35	1.5	6.3	2.57	2	13

Source: Surveys conducted by the author in 1979/80

Table 39. Conditions of land tenure in the three investigation areas, in % of farms

Province	Nueva Ecija	Camarines Sur	Iloilo	$\bar{x}$
Owners	9.6	21.5	12.5	14.5
Leasehold contracts	86.8	69.0	65.0	73.9
Share croppers	1.9	9.5	22.5	11.3

Source: Author's surveys in 1979/80

Average family sizes, on the other hand, differ only marginally, which in turn results in relatively clear differences in the land-man ratios. The cropping acreage per member of family per cropping season amounts to 0.33 ha in Nueva Ecija, 0.22 ha in Camarines Sur and 0.21 ha in Iloilo. The three investigation areas consequently differ with respect to current production conditions.

Owing to the influence of the different yield and productivity levels, the above structural parameters give rise to different socio-economic conditions.

4.2.4.2 Land Tenure

The conditions of land tenure were also found to differ in the three investigation areas of Nueva Ecija, Camarines Sur and Iloilo. However, as Table 39 shows, leasehold contracts predominate. Only about 15% of the farmers interviewed own their farms. These farms are still burdened by amortization payments to state banks under the land transfer scheme, the amounts involved being comparable to rent.

The proportion of share-croppers still left differs greatly in the three investigation areas and is particularly high in Iloilo amounting to 22.5%.

The conditions of land tenure affect the distribution of the incremental yields resulting from pest control measures. In the case of share-cropping, two forms of distribution are customary: in the first case, the landlord receives 50% of the crop and meets 50% of the costs of soil cultivation, transplanting and fertilizing, but rarely the costs of crop protection (Herdt and Jajasuriya 1981). In the second case, the landlord receives only one third of the crop, but does not contribute towards production costs.

The harvest workers' share of production costs is deducted from the overall yields. A farmer share-cropping on a 50% basis will thus only be able to keep about 45% of the incremental yields.

4.2.4.3 Costs and Returns

The yield levels achieved in a specific cultivation area are largely determined by the prevailing natural production conditions and the yield-increasing inputs used. The input levels, on the other hand, are determined both by the natural production conditions and by the institutional and economic background conditions.

Apart from prices, yield and input levels determine the gross margins attained in rice cultivation. The gross margin in turn considerably affect the incomes and the liquidity situation of the farms in a given cultivation area.

Table 40 lists the cost and returns data and the gross margin in rice production for the three areas concerned. The data have been taken from the socio-economic surveys conducted in the 1980 rainy and dry seasons.

It will be noted that the variable costs are higher in the dry season than in the rainy season. This is due to the fact that weather conditions are generally more favourable in the dry season and that higher yields are expected.

The differences are particularly striking in the case of expenditures on fertilizer. If we take into account that fertilizer prices in the 1979/80 dry season were approximately 8% lower than in the 1980 rainy season (Central Bank 1981), we find that between wet and dry seasons there is a relative difference in fertilizer expenditure of as much as 48% in Nueva Ecija and of about 24% in Camarines Sur and Iloilo. With the exception of the extremely high rodenticide input of 35 Pesos ha^{-1} in the dry season in Camarines Sur, as opposed to only 18 Pesos in the rainy season, the differences in expenditure on the remaining inputs are less striking.

Comparing the three cultivation areas, we find that the variable costs in Nueva Ecija are higher than in Iloilo, and lowest in Camarines Sur, amounting to only 2000 Pesos ha^{-1}. The differences in variable costs are mainly caused by expenditures on fertilizer and wage costs. Owing to the alternative possibilities of employment and the more favourable socio-economic conditions existing in the investigation area in the Province of Nueva Ecija, wage and input levels are higher in this area than in the more southern rice cultivation regions which are more distant from the capital.

As regards gross margins, were highest in the investigation area in the Province of Iloilo in the 1980 rainy season. This is mainly attributable to the yield levels achieved in this area.

As shown in Table 41a and b, yield levels differ clearly in the three investigation areas. In Nueva Ecija, considerably higher yields were achieved in the 1980 dry period than in the subsequent rainy season. In Iloilo, on the other hand, the reverse occurred. This was due first to the uncertain water supply during the dry season and second, to the fact that there were no typhoons during the rainy season.

Table 40. Gross margin of rice production in the three investigation areas in 1980 rainy and dry seasons (Pesos ha^{-1})

Season	Cultivation area					
	Nueva Ecija		Camarines Sur		Iloilo	
	Rainy	Dry	Rainy	Dry	Drainy	Dry
Yield (kg)	3329.0	5435.0	3269.0	3756.0	5399.0	4073.0
Price (P kg^{-1})	1.35	1.42	0.98	1.11	1.0	1.24
Returns (P)	4494.2	7717.7	3203.6	4169.2	5399.0	5050.5
Seed	149.8	157.6	155.8	186.2	148.1	128.5
Fertilizer	505.2	691.6	357.5	407.7	379.2	430.7
Insecticides	138.8	119.9	121.1	125.1	132.9	132.8
Herbicides	56.7	46.1	38.4	37.1	54.8	54.6
Rodenticides	21.8	12.8	18.4	35.3	11.9	16.6
Irrigation	141.1	199.0	179.6	103.1	143.2	220.4
Wages, hired lab.	677.9	677.9	588.1	565.5	600.0	588.7
Harvesting costs	637.2	1021.0	345.1	434.0	594.0	756.4
Interest payable	187.6	213.7	143.3	128.8	176.8	183.5
Total variable costs	2516.1	3139.6	1965.3	2022.8	2240.9	2512.2
Gross margin	1978.1	4578.1	1238.3	2146.4	3158.1	2538.3
Rent	469.8	518.4	675.0	787.4	941.5	1069.8
Gr. margin − rent	1508.3	4059.7	563.3	1359.0	2216.6	1468.5
Family labour (man-days ha^{-1})	35	35	30	30	31	31

[a] Calculated on the basis of cash expenditures (fertilizer, crop protection, wages of hired labour), and a composite rate of interest (bank and private loans) for a term of 3 months. For Nueva Ecija, a rate of 13.8% was assumed, for Camarines Sur 11.0% and for Iloilo 15.0%

Source: Author's surveys in 1980

The Province of Camarines Sur ranks last with respect to both yield and expenditure. Also, yield differences between rainy and dry season are lowest. This is attributable to the fact that climatic differences over the year are less pronounced.

4.2.4.4 Potential Yield

The absolute crop losses have to be derived from the potential yield, i.e. from the yield level that would be achieved if pests either do not occur, or if their influence is completely eliminated.

For determining this potential yield which varies from area to area, it is not possible to use the farmers' actual yield levels, since these always reflect a residual pest influence despite the control measures instituted. The only possibility is to use data from trials in which pest influence was virtually eliminated. We can

Table 41a. Maximum yields in kg ha^{-1} obtained in insecticide trials at three experiment stations, 1970–1981 rainy season[a]

Area	Year												$\bar{x}$	Coeff. of variation	t-Values
	1970	1971	1972	1973	1974	1975	1976	1977	1978	1979	1980	1981			
MRRTC (Nueva Ecija)	2150	–	–	4144	–	–	3970	5300	–	4791	4809	4367	4218	24.1	0.69
BRES (Camarines Sur)	–	–	5250	3367	–	4600	–	4636	3710	4260	4529	5330	4460	15.2	0.48
VRES (Iloilo)	5328	4434	3699	–	3494	4992	4265	4218	6254	3897	5268	–	4584	18.8	1.23

[a] Figures are the annual averages of the maximum yields of all trials performed in 1 year

Source: Annual reports of the MRRTC, BRES and VRES stations 1970–1981

Table 41b. Maximum yields in kg ha^{-1} obtained in insecticide trials at three experimental stations, 1975–1981 dry season[a]

Area	Year							$\bar{x}$	Coeff. of variation	t-Values
	1975	1976	1977	1978	1979	1980	1981			
MRRTC (Nueva Ecija)	6836	–	4185	–	6455	4650	–	5531	23.7	0.32
BRES (Camarines Sur)	4125	–	5390	5273	5720	4260	4122	4815	15.0	0.15
VRES (Iloilo)	–	–	6336	5600	3490	4400	–	4956	25.4	0.18

[a] See Table 41a

Source: Annual reports of the MRRTC, BRES and VRES stations 1975–1981

Table 42. Yield levels actually achieved by farmers in the three investigation areas (kg ha^{-1})

Parameter	Province season					
	Nueva Ecija		Camarines Sur		Iloilo	
	Rainy	Dry	Rainy	Dry	Rainy	Dry
$\bar{x}$	2950	3836	3064	3465	4424	3124
Coefficient of variation	31.1	19.5	6.5	4.7	6.94	25.03
Min.	1900	2400	2890	3265	4210	2636
Max.	4150	4800	3270	3625	4937	4720

Source: Bureau of Agricultural Extension (1972–79)

therefore fall back once more on the pesticide evaluation trials, employing the yield from each trial as potential yield.

To check whether the yield data recorded at experimental stations correspond to actual field values, they can be compared with long-term yield records such as those kept within the scope of the national rice production program.

Table 42a and b give the mean maximum yields of all the insecticide trials performed per year, broken down according to trial site and cultivation period. Since trial results are not available for every year (since 1970), some of the mean values are based only on four observations. Disregarding these unfavourable statistical conditions, we can say that no significant yield differences between the three areas can be ascertained during the rainy season. The appropriate *t*-values for the individual values on which the means are based, are listed in Table 41a and b. However, at the MRRTC station, yield variance (expressed in terms of the coefficient of variance), amounting to 24.1, is clearly higher than at the other two experiment stations.

In the dry season, the MRRTC station in the Province of Nueva Ecija has the highest average yield level of 5531 kg ha^{-1}.

As already mentioned earlier in connection with the yield-cost structure, the water supply in the Province of Iloilo is uncertain during the dry season. This is reflected in a coefficient of variation in the mean yield of 25.4%.

Comparing the maximum yields obtained in the insecticide trials with the average crop yields actually achieved in the irrigated rice cultivation areas in the different provinces, we find that the rainy season yields are fairly similar (Table 42). During the dry season, on the other hand, greater differences can be observed. This is probably attributable to the higher fertilizer inputs – based on the national recommendations – and the better water supply at the experiment station.

4.2.4.5 *Crop Prices*

Within the scope of the production promotion program, the price of rice was fixed by the Government and amounted to 1.5 Pesos per kg in 1980. However, only a small proportion of the farmers sell their crop to the Government buying

Table 43. Nominal prices of rice in Pesos kg^{-1}, and cost of living index, 1975–1980

	Central Luzon	Bicol	Western Visayas	Philippines overall	Cost of living index
1975	1.0	0.88	0.86	0.92	100
1976	1.08	0.97	0.93	0.96	109
1977	1.13	0.99	0.92	1.0	120
1978	1.07	0.96	0.90	1.01	128
1979	1.05	0.95	0.97	0.99	150
1980	1.19	0.99	1.06	1.08	177

Source: Bureau of Agricultural Economics and Central Bank of the Philippines, 1980

centres, since these make high demands on quality and do not have sufficient transport facilities.

According to the survey, 73% of the farmers in Iloilo and 84% in Nueva Ecija sell their products to private dealers. The main reasons given by the farmers are prompt payment (56%), better transport facilities (10%) and also prices (34%). The latter is due to the fact that government buying centres apply price reductions if quality standards are not being reached. The price actually attained by the farmers is clearly lower than the government minimum prices, with considerable regional and seasonal fluctuations.

The highest average annual price of 1.19 Pesos was observed for the Central Luzon region in which the Province of Nueva Ecija is located, whilst the Bicol region with the Province of Camarines Sur reached only an annual average of 0.99 Pesos kg^{-1} (Table 43).

The monthly price fluctuactions are connected with the peak harvest times. Prices are generally highest from the beginning of the planting season until the beginning of the harvest period. Price fluctuations during the main harvest periods in September/October (rainy season) and in February/March (dry season) are lower. The price in Central Luzon was about 10% higher in October than in April, the main harvest month in the dry season.

It is impossible within the scope of the present study to elaborate a sufficiently accurate price forecast for determining control thresholds and for selecting the optimum crop protection strategy. However, decisive changes in product prices are not likely to occur, so that from this angle the economic feasibility of crop protection measures is not likely to be affected to any marked degree.

Studying the price development at national level since 1975 (Table 43), we find that the nominal ex-farmgate prices of rice have increased only marginally. A comparison of this price development with the change in the cost of living index shows that in real terms rice prices have even dropped.

For the model computations, a breakdown of rice prices according to rainy and dry season will be necessary.

Table 44. Disposal of the rice harvest

Investigation area	Crop sold	Payment of harvest workers	Rent	Farm household consumption	Loan repayment	Irrigation charges	Seed requirements
Nueva Ecija	24.6	13.7	8.6	12.8	32.9	3.9	3.5
Camarines Sur	19.8	10.5	20.0	21.3	21.4	4.1	2.9
Iloilo	33.7	14.2	19.3	16.2	9.2	4.0	3.3
$\bar{x}$	26.0	12.8	16.0	16.8	21.2	4.0	3.2

Source: Surveys conducted by the author in 1980

4.2.4.6 Farm Income Situation and Necessary Minimum Yields

In assessing the net farm income-situation, the close connection existing between farm and farm household must be taken into account. Since rice is the staple food, a considerable part of the output is used either for the farm household, for loan repayment or for the payment of other taxes.

Table 44 shows the average utilization of the rice output in the three investigation areas. Furthermore, it can be seen that only about 25% of the rice produced is sold and that almost 30% is deducted for payment of rent and harvest workers. On the average, over 20% of the rice production is needed to repay short-term loans, i.e. money borrowed to finance the purchase of inputs and consumer goods up to harvest time. There are considerable regional differences which are probably attributable to the differences in economic background conditions and consumer behaviour. As regards the remainder of the output, an average of 17% is used for the farm household, and 4 and 3.2% respectively for the payment of irrigation charges and seeds.

The receipts available to the farm from the sale of rice is supplemented by external, partly non-farm incomes. External income is earned by the farmers and their families by working additionally on neighbouring farms where crops are grown earlier or later. In some regions, temporary work on sugar cane plantations is a further external source of income. The non-farm income consist of contributions by members of the family engaged in non-farm work. To cover household expenditures – including, as a major cost item, expenditure on the education of the children – a varying proportion of the rice production is needed. This proportion depends on the level of household expenditures, the size of the outside income, yield levels and crop prices.

The data on household expenditures, as well as off and non-farm income obtained from the farm surveys conducted in the three investigation areas are shown in Table 45. The figures computed apply to average farm and family sizes and refer to 1 ha.

Household data were covered only once in the surveys, so that the same amount has been entered for rainy and dry seasons alike. For computing the

Table 45. Proportion of rice output ha^{-1} needed to cover consumption expenditures of the farm household, 1980 rainy and dry season

	Nueva Ecija		Camarines Sur		Iloilo	
	RS	DS	RS	DS	RS	DS
Consumption expenditure without rice in P ha^{-1}	1889	1889	1577	1577	2625	2625
Non-farm income	1056	720	200	407	968	896
Difference (P ha^{-1})	833	1163	1377	1170	1657	1729
Difference (kg ha^{-1})[a]	617	823	1405	1054	1657	1394
Difference as a percentage of yield[b]	18.5	15.1	42.9	28.1	30.1	34.2
$\bar{x}$ rainy and dry seasons	16.8		35.5		32.2	
Non-farm income as a percentage of yield	23.5	9.3	6.2	9.8	17.9	17.7
$\bar{x}$ rainy and dry seasons	16.4		8.0		17.8	

[a] Difference in Pesos divided by the price per kg (cf. Table 40)
[b] Difference in kg on the basis of yield (cf. Table 40)
Source: Based on author's surveys in 1979 and 1980

proportion of the rice output needed to cover household consumption, the price and yield data listed in Table 40 were used.

As can be seen in Table 45, depending on cropping period and investigation area, 15 to 43% of the rice output has to be used to cover private consumption.

On the basis of these computations we can now determine the range of variation of the minimum yield – expressed as a percentage of the yield which is expected on average (potential yield) – which will figure as a restriction in our model assumptions.

The lower limit of this range covers the payment in kind of harvest workers and rent, the farm-household consumption of rice and repayment of short-term loans. To the remaining portion of the yield we must add the non-farm income expressed, as before, as a percentage of crop yields.

In this way we obtain the lower limit of the absolute minimum yields for the three investigation areas. A crop protection strategy falling short of these computed limits (constraints) in only one infestation event would be ruled out as an alternative. Depending on the cultivation area concerned, the values listed in the table represent 40 to 65% of the yields of a "normal" year. The corresponding upper limit of the range of fluctuation is equal to the lower limit plus the proportion of household consumption to be financed from the rice crop. As shown in Table 46, the upper limit for the rainy season amounts to almost 2950 kg ha^{-1} for Nueva Ecija, to about 2500 kg ha^{-1} for Camarines Sur and to over 1900 kg ha^{-1} for Iloilo. In Nueva Ecija, the corresponding figures for the dry season are higher, whilst in the other two areas they are lower. This is due to the difference in prices and in the availability of additional sources of income. However, as re-

Table 46. Upper and lower limit of minimum yield (kg ha^{-1})

	Nueva Ecija		Camarines Sur		Iloilo	
	RS	DS[a]	RS	DS	RS	DS
Harvest workers' share	472	719	345	391	722	610
+ Rent	348	365	689	709	941	868
+ Farm household consumption of rice	507	567	664	834	768	748
+ Short-term loan	1004	1934	1057	394	469	393
– Non-farm income in kg	782	496	204	367	968	723
Lower limit	1549	3089	2551	1961	1932	1896
+ Consumption expenditure (in kg of rice)	1399	1303	1609	1421	2625	2116
Upper limit	2948	4392	4160	3382	4557	4012

[a] RS = rainy season, DS = dry season
Source: Computed by the author on the basis of surveys, 1979 and 1980

gards the latter, the persons interviewed showed a certain reluctance to be specific about these incomes. It is nevertheless clear that the income situation in the investigation area in the Province of Camarines Sur is the most unfavourable one. This is also evident if we consider the gross margins (Table 40).

The computed limit values, which figure as restrictions in the model computations, represent a behaviour assumption which has not been verified by the author by means of empirical surveys. For this reason, the returns from strategies which result in favourable average net returns and fall marginally short of the above minimum yields in only one or a few cases, will also be covered in the interpretation of results.

4.2.4.7 Liquidity Position

The detailed account given above of the source and utilization of the net incomes in the investigation areas also provides information on the liquidity position of the farms. Under the given conditions, the limited availability of cash funds acts as a restriction with respect to the selection of crop protection strategies which are more costly than the measures currently used by the farmers.

As shown in Table 47, hardly any cash surplus remains when considering the present level of household consumption. However, the results of our computations for Nueva Ecija, which are based on 1980 data, do not seem to be representative because of the comparatively high crop prices (cf. Table 40).

At this point it should be mentioned that income data obtained by means of surveys tend to be unreliable. The figures listed in the preceding tables are the averages of the samples concerned. They should, therefore, be considered only as an indication of the order or magnitude involved.

Table 47. Cash surplus and deficit in Pesos ha^{-1}, 1980 rainy and dry season

	Nueva Ecija		Camarines Sur		Iloilo	
	Rainy season	Dry season	Rainy season	Dry season	Rainy season	Dry season
Cash income	1105	1898	634	825	1819	1702
+ Non-farm income	1056	720	200	407	968	896
− Consumption expenditures (excl. rice)	1889	1889	1577	1577	2625	2625
Cash surplus/deficit	+272	+729	−743	−345	+165	−27
Long-term debts	404	325	525	459	304	439

[a] Computed from the gross returns ha^{-1} and the proportion of the rice output sold (Tables 40 and 45)

Source: Computed from author's surveys, 1980

With regard to the liquidity position of the farms, it is nevertheless possible, on the basis of these surveys, to draw the conclusion that the farms would hardly be in a position to increase their input level of one input factor without correspondingly reducing that of another. The long-term liabilities listed in Table 47 also underline this fact. In the light of further price increases of production inputs coupled with a simultaneous decline in real terms of crop prices and a less pronounced rise in yield levels, the liquidity position of the farms can be expected to deteriorate rather than improve in future.

4.2.4.8 Opportunity Costs of Cash

If we were to assume that the farmers have exhausted all their cash resources, we would have to take into account the yield of the next best alternative for computing the cost of using one input factor, i.e. the use of cash would involve opportunity costs. Applied to the field of crop protection, this means that every additional unit of money expended on crop protection inputs corresponds to the non-use of that unit of money for a different input, e.g. fertilizer. This fact is illustrated in Fig. 12. In this purely hypothetical presentation, it is assumed that there is a functional relationship between cash inputs and yields. Depending on the infestation situation concerned and on whether or not crop protection measures are instituted, three different production functions have to be considered. To begin with, we will select function A, representing a situation without pest infestation. In accordance with his available cash funds, the farmer would aim at input level F_2 of yield increasing input factors (e.g. fertilizer) and obtain yield EA A. If, contrary to all expectations, a pest infestation does occur, then with fertilizer input remaining constant, production will take the course of function B and yield EB A will result. If it had been known from the beginning that an infestation would occur, the farmer could have used part of his funds for crop protection and reduced his fertilizer input to level F_1. Function C would then have been applied and yield EC C would have been achieved. Vice versa, the farmer

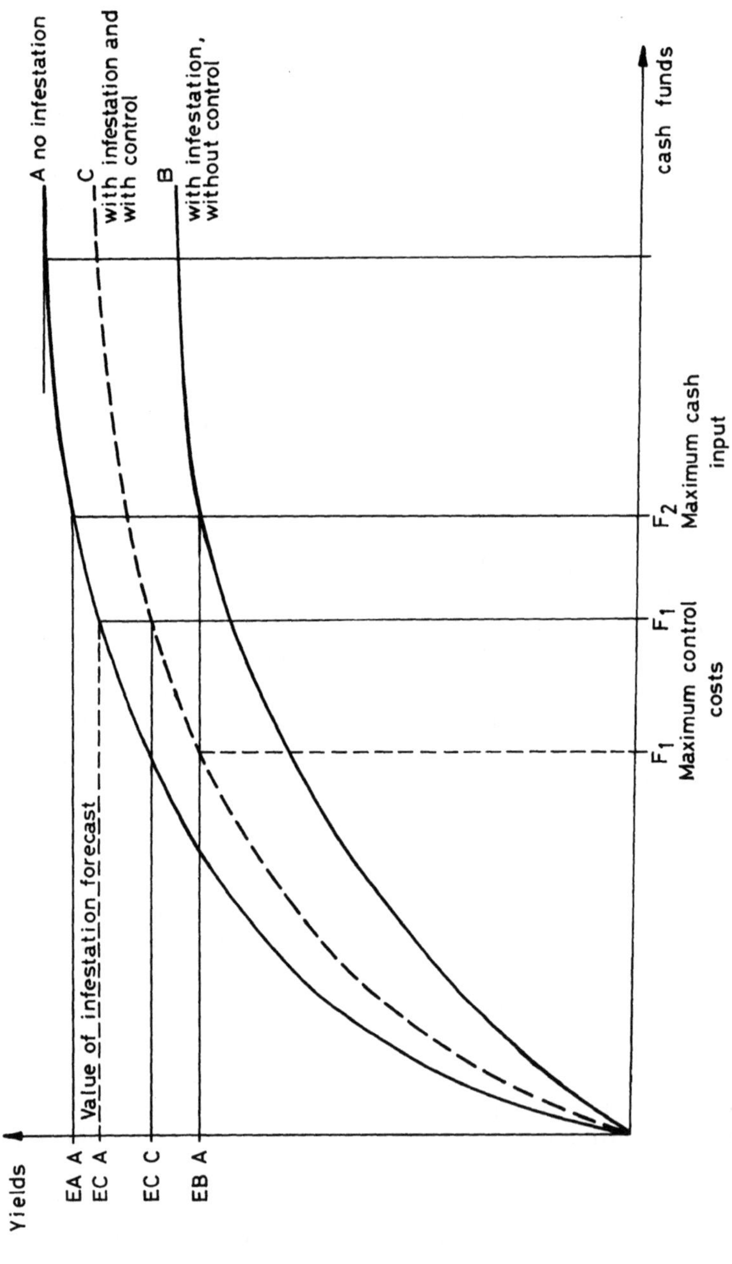

Fig. 12. Impact of pest control on yield if the availability of cash is limited

could always prepare for function C and if an infestation then occurs, his level of inputs would be optimum. On the other hand, if no infestation occurs, the net returns from the crop protection measure would be equal to zero and the fertilizer input would be below the optimum level. Instead of the possible yield EA A, only yield EC A would be realized. The difference between these two yield levels consequently corresponds to the opportunity costs of crop protection

Table 48. Production functions of rice cultivation in the three investigation areas

Area (province)	Year	Season	Type of function (*t*-values)	Exogenous variables[a]	R^{2b} (r)	*F*-value	n^c
Nueva Ecija	1980	Rainy and dry seasons	$Y = 4262 - 1961x_1^{***} + 1022x_2^{**} + 9.01x_3^{*}$ (7.84) (4.1) (2.2)	x_1 = Dummy variable for cropping season[d] x_2 = Dummy variable for K_2O[e] x_3 = kg N	0.518	31.58**	92
Iloilo	1980	Rainy season	$Y = 3748 + 58.98x_1^{*} + 56.98x_2^{*}$ (2.23) (2.17)	x_1 = Herbicide input in Pesos x_2 = Weeding in man-days	0.23	5.56*	39
	1979	Dry season	$Y = 2807 + 12.209x_1^{*} + 35.28x_2^{*}$ (2.82) (2.33)	x_1 = Herbicide input in Pesos x_2 = Weeding in man-days	0.25	6.18*	39
	1980	Rainy and dry seasons	$Y = 3415.7 + 1264x_1^{**} + 12.14x_2^{*}$ (5.62) (3.22)	x_1 = Dummy variable for cropping season 0 = dry s., 1 = rainy s. x_2 = Nitrogen input	0.34	19.58**	78
Camarines Sur	1980	Rainy and dry seasons	$Y = 3677 + 6108x_1^{*}$ (2.64)	x_1 = Herbicide input in Pesos	0.36	6.97*	48

[a] The endogenous variable used was always yield in kg ha^{-1}
[b] Coefficient of correlation when there is only one exogenous variable
[c] No. of observations (farms)
[d] 0 = dry season, 1 = rainy season
[e] 0 = without K_2O, 1 = with K_2O
* Significant at $P \leq 0.05$
** Significant at $P \leq 0.01$
*** Significant at $P \leq 0.001$
Source: Author's computations

in the case of an infestation intensity below the economic damage threshold. From this value we can also compute the benefit to be derived from basing a strategy on control thresholds, if we assume that control effectiveness will be the same as that of prophylactic measures when the control threshold is reached. To obtain an idea of the order of magnitude of the opportunity costs of cash in the three investigation areas, it will be necessary to examine the various production alternatives, in particular fertilizer input and weed control.

For this purpose, production functions were constructed on the basis of the socio-economic data. For the Nueva Ecija area, the marginal yield resulting from N fertilizer inputs was computed as 9.01 kg of rice kg^{-1} N (Table 48). For computing the marginal yield resulting from K_2O inputs, a dummy variable had to be used since only some of the farms had applied potash. The average input level of 20.2 kg of K_2O ha^{-1} used by the farms applying potassium resulted in a highly significant incremental yield of 1,022 kg ha^{-1}.

In Iloilo, it was possible to prove the positive influence of both weed control and application of N fertilizer, the marginal yields of weed control being considerably higher than those of N fertilization. However, the test values of the weed control functions based on the 1979 data were found to be considerably lower than those of the N function for 1980 (Table 48). For this reason it appears to be more realistic to employ the marginal yield resulting from N fertilizer inputs as the alternative use of cash.

In Camarines Sur, statistically significant proof of an effect on yield was obtained only in the case of herbicide inputs. Despite the uncertainty involved, a marginal yield of 6.1 kg of rice was therefore assumed for each additional unit of money invested in weed control.

The above computations regarding the actual on-farm value of cash deal only with the use of inputs in rice cultivation. For investing cash savings, there may well be other possibilities with marginal yields exceeding those obtained in rice cultivation. However, since only scant information is available in this regard, alternative crop or livestock enterprises, will not be dealt with in the present study.

In this context, application of the principle of opportunity costs will always be based on the assumption that any cash funds saved are reinvested in the farm and not used for consumption purposes.

4.3 Results of the Model Computations

From model computations we can derive a number of guidance criteria for crop protection decisions.

First of all, we will have to identify the crop protection strategy (strategies used are defined in Tables A1–A4 of the Appendix) which can be regarded as optimal for the given objective, both for situations in which resistant varieties are available and for cases in which varietal resistance to the main damage factors (e.g. tungro virus, brown planthopper) has broken down.

To determine this optimum strategy, we will first of all compute the net returns (value of output less costs of control) of the crop protection strategies described in Section 4.2 for all the infestation events (e.g. year and season, always at the foot of the following tables) in resistant and non-resistant varieties discussed in Chapter 3. From this pay off matrix, using the selection criteria listed in Section 4.2, we will then select the optimum strategy. The net returns of the optimum strategy will serve as a measure of comparison for the net returns to be derived from alternative strategies, e.g. from the various strategies used by the farmers or currently recommended by the extension service.

In view of the different preconditions regarding infestation events and yield, price and cost parameters prevailing in the three investigation areas and in the rainy and dry seasons, the optimum strategy will be determined separately for the three areas and for rainy and dry periods.

The model computations will also yield data for determining the benefits to be derived from the provision of information designed to improve the farmers' basis for decision-making on crop protection strategies.

The first level consists of the warnings issued by the official crop protection service. These warnings are intended to inform farmers at an early stage of impending pest outbreaks or of the breaking down of varietal resistance.

This information can also take the form of negative prognosis indicating the probable durability of varietal resistance. For the decision-maker (farmer), this limits the decision problem to the events in either resistant (E_1) or non-resistant varieties (E_2). The benefits of the warning will thus correspond to the difference between the net returns to the optimum strategy in the case of the combined events in resistant and non-resistant varieties (E_1 and E_2) occurring, and the net returns to the optimum strategy in the case of one of the two sets of events occurring.

The second level of information improvement consists of the use of control thresholds. Systematic pest monitoring will make decisions possible on chemical control measures to be taken on the basis of control thresholds, the pesticides used being those recommended by the official extension service (cf. Sect. 4.2.1.4). As a result of the low level of effectiveness of the control thresholds currently recommended (cf. Sect. 4.2.2.4), the main benefits of their application lies in savings in control costs. If production alternatives exist, the benefit of control thresholds can be computed from the yields resulting from the alternative use of the cash funds saved.

The final level of information is arrived at when reliable infestation prognoses are available. This level of information can also be regarded as attained if, irregardless of whether or not the threshold is reached, the use of control thresholds leads to net returns unexceeded by any other control strategy. However, in view of the present state of knowledge, these computations will merely serve the purpose of deriving the potential benefits of short-term infestation prognoses. The potential benefits are computed by comparing the average maximum net returns obtained for the different infestation events with those from the optimum strategy. Thereby neither infestation forecasts nor control thresholds are involved and the marginal returns of any cash funds possibly saved are taken into account.

In the following, first the results computed on the basis of current background conditions will be presented. Subsequently, the effects of changes in important influence parameters will be studied.

Table 49. Matrix of infestation events in non-resistant varieties, Nueva Ecija 1978–1981

Pest	Infestation event[a]												
	1	2	3	4	5	6	7	8	9	10	11	12	13
Rice whorl maggot[b]	–	–	–	–	–	–	–	–	–	–	–	–	2.4
Army worm with rice caseworm + other leaf feeding pests (generative phase)	–	–	–	28.9	–	5.5	–	2.81	10.3	5.0	2.39	3.0	25.2
Army worm[b] (generative phase)	–	–	–	–	–	–	–	–	5.1	–	–	–	2.0
Rice leaf folder[b] (generative phase)	35.0	8.0	31.0	7.6	16.0	2.4	7.6	6.0	–	4.8		19.8	–
Stem borer[c] (generative phase)	–	12.5	7.0	7.4	–	–	–	–	2.5	–	2.61	6.2	6.25
Stem borer[c] (ripening phase)	7.2	–	–	–	1.8	3.4	–	–	1.7	7.2	3.61	2.7	7.09
Year	1978	1979	1979	1979	1980	1980	1980	1980	1980	1981	1981	1981	1981
Season[d]	RS	DS	RS	RS	DS	DS	RS	RS	RS	DS	DS	RS	RS
Source[e]	A 1980 (57)	A 1980 (58)	A 1981 (61)	Un-published	A 1980 (59)	A 1982 (80)	A 1980 (60)	1980 (85)	B	A 1981 (55)	B	1981 (59)	B

[a] See year and season below
[b] Percentage of infested leaves
[c] Percentage of dead hearts/white panicles
[d] RS = rainy season, DS = dry season
[e] A = IRRI – Insecticide Evaluation Report, figures in brackets = nos. of tables; B = author's trials

4.3.1 Results Computed for the Nueva Ecija Investigation Area

4.3.1.1 Model Assumptions

The infestation events used in our model computations occurred partly on the untreated plots of on-farm trials carried out on the fields of farmers in the vicinity of the MRRTC, using resistant varieties IR-36, IR-42, IR-50 and IR-54. The events in non-resistant varieties occurred on the untreated plots of insecticide trials conducted with the IR-22 variety.

Table 49 shows the infestation events of resistant varieties in a matrix form. Since a significant difference could not be shown to exist between dry season and rainy season events, (cf. Chap. 2.2.6.1.2, Table 23), the two have been combined. Studying the infestation data, it will be noted that stem borer and rice leaf folder occur most frequently. The rice leaf folder exceeds its threshold of 15% infested leaves in three cases and the stem borer its threshold of 5% dead hearts in four cases, with both pests transgressing their respective thresholds in one instance. In addition, other leaf pests (mainly the army worm) reach their control thresholds of 25% infested leaves in two instances. In almost 50% of the cases, three or more pests occur jointly. For the empty boxes in the matrix of infestation events, an infestation intensity of zero has been assumed, since the trial results did not contain any data on the incidence of these damage factors.

The corresponding matrix for non-resistant varieties (Table 50) is limited to tungro virus and brown planthopper. Here again no differences could be shown to exist between rainy and dry season (cf. Chap 2.3.7.1.1, Tables 21a–c) and the two sets of data were therefore combined. The infestation data listed are the averages of all the trials of 1 year.

The infestation situations are characterized by tungro incidence; a high level of attack by the brown planthopper occurred only in 1977.

The economic parameters assumed for the model computation are listed in Table 51. For crop prices, the regional averages for the months of May and September have been used (cf. Chap 4.2.4.5). To obtain the marginal yields resulting from the cash savings, the regression coefficient of the fertilizer function in Table 48, amounting to 9.01 kg of rice per kg of N, was related to one unit of money (Peso), a nitrogen price of 4.58 Pesos per kg of N being assumed. This results in an incremental yield of 1.97 kg of rice for each additional unit of money spent on N fertilizer. Relating this figure to the potential yield in the

Table 50. Matrix of infestation events in non-resistant varieties, Nueva Ecija 1973–1981

Pest	Infestation event							
	1	2	3	4	5	6	7	8
Tungro virus[a]	9.8	49.75	86.53	7.03	36.3	0.96	2.89	26.77
Brown planthopper[b]	0	0	0	72.2	0	7.5	22.89	22.40
Year	1973	1975	1976	1977	1978	1979	1980	1981

[a] Infested hills in % during the generative growth stages
[b] No. of planthoppers per hill
Source: MRRTC – Annual Reports, Entomology Division

Table 51. Economic parameters used in the model computations for Nueva Ecija

Cropping season	Crop price (P kg^{-1})	Interest per ropping period (%)	Marginal yield of cash as a percentage of the yield per Peso	Potential yield (kg ha^{-1})	Financial limit (P ha^{-1})
Rainy season	1.27	13.8	0.047	4218	240
Dry season	1.14	13.8	0.036	5531	240

Source: See Chapter 4.2.4

Table 52. Yield achieved by means of strategies S15 and S45 in non-resistant varieties in kg ha^{-1}

Strategy[a]	Season[b]	Infestation event							
		1	2	3	4	5	6	7	8
S15	RS	3997	3720	3661	2911	4163	4327	4118	1415
	DS	5202	4842	4765	3789	5419	5632	5361	1842
S45	RS	4113	4044	4029	3312	4156	4208	4076	1844
	DS	5392	5303	5284	4343	5449	5519	5345	2418

[a] See Tables A1 and A3 of the Appendix
[b] RS = rainy season, DS = dry season

rainy and dry season respectively, we obtain the values listed in Table 52. The rate of interest shown refers to one cropping period and corresponds to the rates given in Section 4.2.3.3.

A cash constraint of 240 Pesos ha^{-1} was assumed for the implementation of crop protection measures. This corresponds to the cost of the most expensive strategy used by farmers in the Nueva Ecija area.

4.3.1.2 The Optimum Strategy

The complete matrix for the model computations based on the conditions currently prevailing in the investigation areas covers a total of 1,470 values resulting from 13 infestation events in resistant varieties, eight events in non-resistant varieties and 70 control strategies. Since without the information provided by the early warning service, an infestation situation in resistant varieties as well as a breakdown of varietal resistance has always to be reckoned with, both types of events were regarded as being equally likely. The optimum strategy will consequently be that, which, conforming to the assumed liquidity and minimum yield restrictions, will on the average – considering both infestation situations – result in the highest net returns.

Table 53 lists the net returns from the best strategies within the individual groups of measures. Under the preconditions mentioned above, strategy S 15 is the best for both the rainy and the dry seasons. The strategy concerned consists

Table 53. Average net returns from alternative crop protection strategies in Pesos ha^{-1}, resistant and non-resistant varieties, rainy and dry season

Strategy[a]	Control costs (P ha^{-1})	Rainy season					Dry season				
		With opportunity cost of cash			Without opportunity cost of cash		With opportunity cost of cash			Without opportunity cost of cash	
		Net returns (P ha^{-1})	Min. yield[b] (kg ha^{-1})	Marg. yield[c] (%)	Net returns (P ha^{-1})	Min. yield[b] (kg ha^{-1})	Net returns (P ha^{-1})	Min. yield[b] (kg ha^{-1})	Marg. yield[c] (%)	Net returns (P ha^{-1})	Min. yield[b] (kg ha^{-1})
S 5	223.5	4077.6	1070	0.67	4064.9	1063	4834	1402	0.51	4824.2	1395
S 6	91.3	4327.8	1127	6.02	4212.4	1063	5063.8	1459	4.61	4974.4	1395
S10	238.4	4151.1	1063	0.06	4150.0	1063	4927.8	1395	0.05	4927.2	1395
S11	238.4	3606.2	276	0.06	3605.2	276	4286.5	362	0.05	4285.8	362
S12	158.2	4294.6	1098	3.31	4228.1	1063	5056.7	1430	2.54	5004.7	1395
S13	237.3	4170.5	1063	0.11	4168.2	1063	4950.0	1395	0.08	4948.3	1395
S14	237.3	3548.8	276	0.11	3547.5	276	4218.7	1395	0.08	4217.6	362
S15	159.3	4388.5	1415	3.27	4319.4	1370	5166.7	1842	2.5	5112.9	1797
S16	313.0	4244.7	1566	0.0	4244.7	1566	5051.7	2055	0.0	5051.7	2055
S17	159.3	4200.8	1350	3.27	4138.4	1307	4947.6	1757	2.5	4899.4	1714
S23	179.8	4292.5	1542	2.44	4242.6	1505	5064.7	2011	1.87	5025.5	1974
S29	314.0	4276.5	1605	0	4276.5	1605	5089.2	2105	0	5089.2	2103
S36	295.9	4280.4	1736	0	4280.4	1736	5090.7	2275	0	5090.6	2275
S38	248.1	3710.1	276	0	3710.1	276	4410.9	1469	0	4410.9	1469
S45	627.9	4170.8	1844	0	4080.4	1844	5020.5	2418	0	5020.5	2418
Non-control	0	3774.9	303	9.72	3659.1	276	4387.8	389	7.45	4307.0	362

[a] See Appendix
[b] Minimum yield = lowest yield realized
[c] Based on yield level achieved

of the application of Carbofuran in granular form before transplanting. The cost of this crop protection measure amounts to about 160 Pesos ha^{-1} and can be financed by the majority of the farms. Taking into account all infestation events, the average return in the rainy season obtained from routine prophylactic applications of Carbofuran amounts to 4388 Pesos ha^{-1}, including an incremental yield of 3.27% resulting from the remaining cash funds of 80 Pesos ha^{-1} ($240 - 160 = 80$) (Table 51). If we assume that the cash funds saved are not put to any productive use, the net return will amount to only 4319 Pesos ha^{-1}. The marginal net return obtained from crop protection in the rainy season can be derived from the difference between the net return in the optimum strategy and the net return in the case of non-control. Taking into account the marginal yield of the cash funds saved, this results in 613 Pesos ha^{-1}, disregarding the latter, it will result in 660 Pesos ha^{-1}. If the opportunity costs of cash are taken into account, the economic feasibility of crop protection measures always decreases, since in situations with low infestation levels, the maximum relative yield increment of 9.72% results in non-control being superior.

For the dry season, the same infestation events were assumed, but different economic parameters were used (Table 51). Here, too, S 15 is the most favourable control strategy. The net return to be derived from its application amounts to 5166 and 5112 Pesos ha^{-1}, the difference between control and non-control amounting to 779 and 805 Pesos ha^{-1} respectively. The difference in returns obtained by including or excluding the opportunity costs of cash is smaller in the rainy season, lower marginal yields having been assumed for this period in view of the higher yield levels achieved (Table 53).

It has been shown in Section 4.2 that an economically valid assessment of crop protection measures cannot be based solely on cost and returns considerations. On the contrary, in view of the difficult financial situation of the small rice farms, it is necessary to examine the results of these strategies under conditions of extremely high infestation. For this reason, the minimum yields of all the strategies examined for all the infestation events considered are listed in Table 47. These data now have to be compared with the assumed minimum yield of 1549 kg ha^{-1} in the rainy season (Table 46). It will be noted that strategy S 15, with a minimum yield of 1415 kg, falls about 130 kg short of this minimum. It is only in the case of strategy S 36, consisting of two early applications of the pesticide Monocrotophos that yields do not fall below 1736 kg. With respect to minimum yields, strategy S 45 with a yield level of 1544 kg ha^{-1} is even better.

For the dry season in view of higher production costs and lower off-farm income, the minimum yield was computed as 3089 kg (Table 46). None of the strategies conforms to this restriction. The highest minimum yield of 2418 kg ha^{-1} is obtained by means of four applications of Azinophosethyl (S 45), whilst the strategy with the highest average net returns (S 15), results in minimum yields of only 1842 and 1797 kg (Table 53).

Although, as mentioned above, the absolute minimum yields assumed in our computations are merely intended to indicate the order of magnitude of these restrictions, it is evident that the rainy season yields would only limit if costly crop protection measures are undertaken. The costs of strategy S 45, for instance, amount to over 600 Pesos ha^{-1}. Owing to the limited cash resources of the farmers, this sum could only be made available if the application of fertilizer were to be almost completely dispensed with. This strategy could therefore be financed solely by means of extensive government subsidies. However, at this point we must consider whether more importance is to be attached to the realization of the minimum yields (or minimization of the loss in yield) or to the magnitude of the average net returns. In the case of S 15, for instance, the average net returns in the rainy season are almost 220 Pesos ha^{-1} higher than those from S 45; in the dry season, they are about 1450 Pesos ha^{-1} higher. On the other hand, S 15 falls short of the set yield minimum only once in the rainy and once in the dry season, although the loss in yield in the dry period, amounting to 1200 kg ha^{-1} is very pronounced.

In view of the above results and on the basis of the assumptions made, we can conclude that the prophylactic application of the pesticide Carbofuran in granular form comes closest to meeting the set objectives. In drawing this conclusion, we are assuming that there is a situation of uncertainty, i.e. that the farmer has no information at all regarding the probability of occurrence of a certain infestation situation and that he must therefore reckon with all possible infestation events, i.e. also with a breakdown of varietal resistance.

If, on the other hand, we employ the maximin rule as a decision rule, then in the rainy season, we would chose strategy S 29. In the infestation event which would lead to the maximum crop loss in the case of non-control (event No. 3, non-resistant varieties), this measure, consisting of two applications of Azinphosethyl at an early growth stage, would result in the maximum net returns of 4279 Pesos ha^{-1}. Yield and liquidity aspects have not been taken into account. Since the infestation events have been assumed to be identical in the dry period, the same will apply with the returns from strategy S 45 amounting to 5042 Pesos ha^{-1} in this case.

Overall it is clear from the above that the strategies used by the farmers do not lead to the maximum net returns – neither using the Laplace rule with its liquidity and minimum yield restrictions, nor applying the maximin rule. As a next step, we will therefore examine the economic feasibility of the farmers' strategies, comparing it with that of the optimum strategy computed above.

4.3.1.2.1 Comparison Between Optimum Strategy and Farmers' Strategies. The farmers do not apply Carbofuran granules in the manner recommended by the extension service. Farms using this pesticide generally apply it shortly after transplanting, which considerably reduces its effect (cf. Chap. 4.2.2.1, Table 33). The full reasons for the farmers' use/non-use of certain pesticides will be given in the final assessment of the results of our model computations for all three investigation areas. At this point, merely the differences in returns will be dealt with.

As shown in Table 53, the farmers' strategy yielding the best result is strategy No. 6 (S 6). However, since the strategy concerned is a control measure of the Camarines Sur area, it is preferable to employ strategy 12 (S 12), being the best farmers' strategy in Nueva Ecija for computing the difference in net returns. Comparing this strategy with the optimum strategy, we arrive at an average difference of 94 Pesos in the rainy season and of 110 Pesos ha^{-1} in the dry season.

Table 54. Basic matrix for computing the benefits of the warning service. Nueva Ecija, rainy and dry season[a]

Strategies[b]	Events			
	Rainy season		Dry season	
	RES (E_1)	NRES (E_1)	RES (E_1)	NRES (E_2)
$SOPT_{nres}$: S29	4457 (4457)	4096 (4096)	5302 (5302)	4876 (4876)
$SOPT_{res}$: S9	5052 (4988)	2500 (2495)	5956 (5904)	2968 (2969)
$SOPT_{res}^{nres}$: S15	4763 (4685)	4014 (3954)	5605 (5542)	4728 (4683)
$SFARM_{res}^{nres}$: S12	4871 (4788)	3718 (3669)	5730 (5663)	4383 (4346)

[a] Figures in brackets refer to the alternative in which opportunity costs of cash are taken into account

[b] $SOPT_{nres}$ = optimum strategy, non-resistant varieties
$SOPT_{res}$ = optimum strategy, resistant varieties
$SOPT_{res}^{nres}$ = optimum strategy, resistant and non-resistant varieties
$SFARM_{res}^{nres}$ = best farmer strategy, resistant and non-resistant varieties

Solely from the point of view of average net returns and without violating the cash constraint, S 12 is only marginally worse than the optimum strategy. However, with respect to minimum yields, there is a difference of 400 kg of rice ha^{-1} in the dry season. This corresponds to as much as 70% of the proportion of the output ha^{-1} computed as farm household consumption (567 kg) for Nueva Ecija.

As regards the effectiveness of the farmers' strategies, it must be pointed out that it is not at all certain whether strategies S 5, S 6, S 10, S 12 and S 13 still correspond to effective tungro control. If we were to assume that these measures do not prove effective – a matter to be expected particularly in the case of high pest density – strategy S 11 would have to be employed for our comparison and a difference in net returns of 782 Pesos in the rainy season and 880 Pesos ha^{-1} in the dry season would be arrived at. At the same time, the minimum yield would decrease below 400 kg, a level no longer acceptable.

On the average, considering all the infestation events to be expected, the farmers' strategies therefore seem to be entirely plausible from an economic point of view. However, when extreme infestation situations occur, their use becomes problematic. In this case, the recommendations of the official extension service would be likely to lead to more favourable results, especially with regard to preventing complete farm illiquidity.

4.3.1.3 Benefit of the Warning Service

In rice cultivation, the main objective of the warning service is to prevent large-scale disease and pest epidemics. Farm-level advising plays a minor role (Heinrichs 1978). The individual farmer is merely informed of the likelihood of varietal resistance continuing or breaking down, this information being based either on measurements of infestation intensity in the fields of selected farms or on the experience of preceding years. If this information is provided early enough for the crop protection strategy to be adapted to it successfully, the farmer's decision problem will be reduced to the events in either resistant or non-resistant varieties. The benefit of the warning service is consequently given by the following equation:

$$N = SOPTE(E_1)p + SOPTE(E_2)1\text{-}p - (SOPT^{nres}_{res}(E_1)p + \mathrm{SOPT}^{\mathrm{nres}}_{\mathrm{res}}(E_2)1-p,$$

where N = benefit of warning issued by warning service in Pesos ha^{-1}

E_1, E_2 = occurrence of possible events in resistant (E_1)/non-resistant (E_2) varieties

$SOPTE(E_1)$ = net returns from optimum strategy for E_1, but E_2 occurs

$SOPTE(E_2)$ = net returns from optimum strategy for E_2, but E_1 occurs

$SOPT^{nres}_{res}(E_1)$ = net returns from optimum strategy for both $E_1 + E_2$ and E_1 occurs

$SOPT^{nres}_{res}(E_2)$ = net returns from optimum strategy for both $E_1 + E_2$ and E_2 occurs

p = reliability of warning issued by the warning service in %.

In computing the benefit, we must distinguish between two different cases. In the first instance, a positive kind of warning is given, i.e. the warning indicates the breakdown of varietal resistance so that the events in non-resistant varieties become relevant. In the reverse case, the prognosis is negative, i.e. a breakdown of resistance to tungro and brown planthopper, the main damage factors, is not likely to occur.

For computing the benefits, Table 54 can be used. To begin with, we will assume that the warnings prove correct. In this case the benefit of a positive type of warning (case 1) is given by the difference between the return from the optimum strategy over both types of events (S 15) and the return from the strategy which is optimum for the situation in non-resistant varieties (S 29). This results in a difference of 82 and 148 Pesos per ha respectively in the dry season. However, if we take into account the cash constraint, it would mean that S 29 could not be realized at all by the farmers. The value of the information provided by the warning service would therefore be equal to zero unless state subsidies are provided or the pesticides are applied by the extension service.

If, instead of S 15, we use the best farmers' strategy as a measure of comparison, in the case of S 29 being implemented, the warning would produce an incremental return of $4095.5-3718.4=377.1$ (rainy season) or $4876.2-4383.1=493.1$ Pesos ha^{-1} (dry season). If the cash constraint is again taken into account,

Table 55. Marginal net returns of the control threshold concept and farmer's strategies in Pesos ha^{-1}, non-resistant varieties, Nueva Ecija

Strategies	Infestation event[a]													
	1	2	3	4	5	6	7	8	9	10	11	12	13	$\bar{x}$
With opportunity cost of cash														
Best farmer strategy	+483.7	−4.5	+385.1	+174.5	+29.0	−213.6	−243.4	−265.7	−37.1	−139.4	−258.8	+208.4	+126.6	+18.7
Control threshold strategy	+187.2 (S61)	−251.2 (S62)	+362.5 (S64)	+31.0 (S66)	−107.7 (S61)	0 (S70)	0 (S70)	0 (S70)	0 (S70)	0 (S70)	0 (S70)	+137.2 (S64)	+22.5 (S66)	+31.2 —
Non-control	4141.7	4676.8	4307.1	4450.3	4874.6	5320.7	5326.1	5351.1	5055.0	4989.4	5421.0	4523.7	4639.6	4852.1
Without opportunity cost of cash														
Best farmer strategy	+554.9	+114.4	+469.2	+274.2	+157.7	−51.0	−79.4	−99.5	+104.1	+1.9	−89.0	+310.7	+237.6	+146.9
Control threshold strategy	+244.4 (S61)	−67.5 (S62)	+525.5 (S64)	+206.7 (S66)	−4.1 (S61)	0 (S70)	0 (S70)	0 (S70)	0 (S70)	0 (S70)	0 (S70)	+319 (S70)	+211.9 (S64)	+110.5 —
Non-control	3993.4	4481.1	4144.1	4274.6	4661.4	5068.0	5072.8	5095.6	4825.9	4766.0	5159.4	4341.9	4447.2	4640.8

[a] From Table 49

then S 15 would be the best realizable strategy. Accordingly, the benefit of the warning service would amount to 4013.9 − 3718.4 = 295.5 Pesos ha^{-1} in the rainy season and to 4728.4 − 4383.1 = 345.3 Pesos ha^{-1} in the dry season. But the implementation of strategy S 15 presupposes that the warning is issued before the rice is transplanted and that it is too late to change to a new resistant variety.

The benefit of the warning service in the case of a negative prognosis can be computed in the same way. First of all, we will again assume that the warning proves correct. If neither a tungro nor a brown planthopper epidemic is going to occur, then the best strategy to adopt would be S 9, producing a return of 5052.8 Pesos ha^{-1} (Table 55). Comparing this strategy with S 15, we obtain a difference in returns of 289.8 Pesos ha^{-1} in the rainy season and 351.5 Pesos ha^{-1} in the dry season. If we assume that S 12 is the initial strategy used, then the difference in returns will decrease to 182 Pesos ha^{-1} in the rainy season and to 226.1 Pesos ha^{-1} in the dry season. Starting out with the farmers' strategy guaranteeing maximum returns over both types of events (S 12), it would thus be possible – in the case of a reliable negative type of forecast – to change to a less cost-intensive strategy. However, if the warning subsequently proves to be incorrect, a definite loss in returns would result (Table 55).

In order to render our computations more realistic, we must abandon our assumption of a warning being absolutely reliable. Owing to the length of the planting period, the great number of varieties used and shortcomings in data collection and transmission, incorrect warnings must always be expected to occur both in time and with respect to a specific cultivation area. If we assume that the warnings are correct 8 times out of 10, then the benefit of a negative type of warning can be computed by substituting this value in the equation given at the beginning of this chapter:

$$N = (5052.8 \times 0.8 + 2469.6 \times 0.2) - (4763.0 \times 0.8 + 4013.9 \times 0.2)$$
$$N = -77.02.$$

The above equation shows that a reliability of 80% of the negative type of forecast would cause the farmer heeding the warning to sustain a loss of over 70 Pesos ha^{-1}. In the case of a positive type of warning of the same degree of reliability, returns would still be achieved:

$$N = (4095.5 \times 0.8 + 4457.4 \times 0.2) - (4013.9 \times 0.8 + 4763.9 \times 0.2)$$
$$N = 3.98.$$

The different figures arrived at are due to the different consequences of incorrect warnings. When mass pest outbreaks are involved, incorrect warnings will not only result in lower returns but also in crop losses, which in the most unfavourable case may lead to farm illiquidity. In the reverse case, an incorrect warning will result in unnecessary additional expenditure on crop protection measures. However, when pest control operations are subsidized by the Government, these costs do not fully affect the farmer.

By solving the two equations for p, we can compute the necessary minimum reliability of a warning. For negative types of prognoses, using S 12 as a measure of comparison, we arrive at a value of 87% reliability for the reverse case, i.e. control recommendations were issued although they were not necessary. This

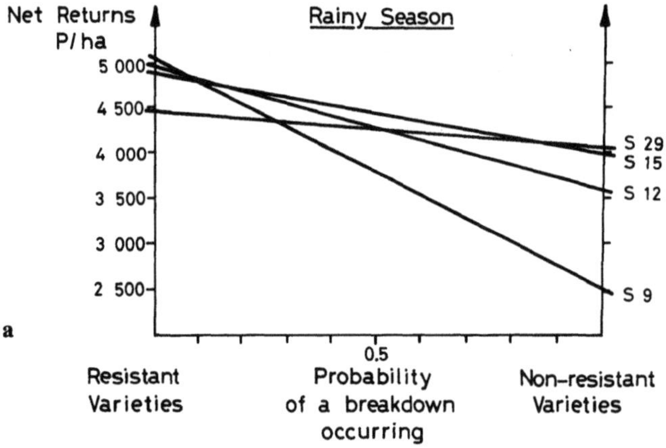

a

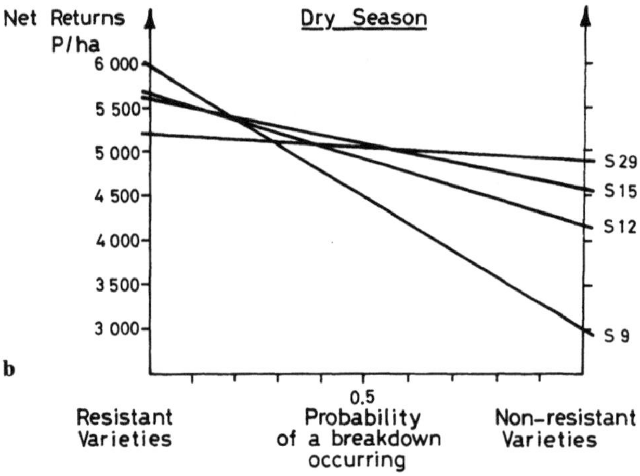

b

Fig. 13a, b. The optimum strategy as related to the probability of varietal resistance breaking down, Nueva Ecija, rainy and dry season

shows that warnings concerning the durability of varietal resistance require a much higher degree of reliability than warnings calling for an intensification of pest control to check or prevent pest outbreaks. This explains why, in actual practice, the warning service issues only warnings concerning probable mass outbreaks.

This fact can also be represented graphically. In Fig. 13a, the net returns from the strategies in a situation with resistant varieties are plotted along the vertical line y_1, whilst the return parameters for a situation in non-resistant varieties are plotted along the vertical y_2. The two verticals refer to the actual occurrence of the two events and indicate the average returns to the strategies weighted according to the probability of occurrence assigned to the events. The point of intersection of two connecting lines always corresponds to the probability at which one strategy becomes economically superior to the other one. This probability in turn corresponds to the required minimum reliability of a warning expressed as a percentage of the number of correct warnings. Fig. 13a, b shows that up to a probability of slightly less than 0.1 of varietal resistance breaking down, strategy

S 9 results in the highest net returns. As the degree of probability increases, strategy S 15 results in the highest returns and it is only when a probability of 0.9 is reached that the use of S 29 would be economically justified.

Fig. 13b illustrates these facts for the dry season. In this case, S 15 does not become superior to the other strategies until a probability of almost 0.2 is reached, while a change to S 29 would be justified at a probability of about 0.7. This demonstrates that the probability of occurrence of a specific infestation situation strongly affects the stability of crop protection strategies. The greater the range over which a specific strategy is superior, the smaller the risk involved in implementing it when an infestation situation has been wrongly assessed. Thus, the application of Carbofuran as a granule prior to transplanting appears to be a rather safe strategy with regard to the set objectives.

4.3.1.4 Benefit of Control Thresholds

The returns to be derived from the use of control thresholds are equal to the average net returns from the corresponding measures taken when the control threshold is reached by one or several pests. The incremental return/loss achieved by means of control thresholds refers only to a situation in resistant varieties, i.e. the basis for the computations is a fully functioning warning service programme. Although control thresholds also exist for the brown planthopper, these cannot be taken into account accurately in our model computations because the damage coefficients have been determined only for the generative growth stage. Account will therefore be taken of these thresholds by equating the brown planthopper's transgression of the threshold with the occurrence of the situation of non-resistant varieties.

As already shown in Table 50 (Sect. 4.3.1.1), the rice leaf folder reaches its control threshold in two cases, the stem borer in one case; in two cases both stem borer and rice leaf folder reach their control thresholds, and likewise, in two further cases, stem borer and army worm. Thus the strategies to be applied are S 61 (twice), S 62, S 64 (twice) and S 66 (twice), and in the remaining six cases, strategy S 70, non-control and the alternative use of cash respectively.

Table 55 lists the net returns to be derived from use of the respective control thresholds, including/excluding the alternative use of the cash funds saved. Our computations are limited to the rainy season since the dry season infestation events are assumed to be identical. Considering all infestation events and taking into account the alternative use of the cash funds saved, the average marginal net returns from the farmers' control strategies amount to only 19 Pesos ha^{-1}. By basing the control measures on the currently recommended control thresholds, a marginal net return of 31 Pesos ha^{-1} compared with non-control, would be achieved. If we neglect the alternative use of cash, then compared with the best farmers' strategy, the average net returns resulting from use of the control thresholds would be 36 Pesos ha^{-1} lower, while non-control would result in an economic loss of as much as 147 Pesos ha^{-1}. If we look at the individual figures arrived at by taking into account the alternative use of cash funds saved, we will note that whenever the control threshold is reached, the farmers' strategy results in net returns higher than those yielded by application of the control threshold concept. Non-control, on the other hand, is always the correct decision when the control threshold is not reached. When the opportunity costs of cash are neglected, the control thresholds are superior to the farmers' strategies in one case each, and the farmers' strategies in turn are superior to non-control also in cases in which the control thresholds are not reached.

Overall, regarding the benefit of control thresholds, we can thus say that in situations in which two pests reach their respective thresholds, two applications at the recommended rates are necessary so that considerable costs are incurred. It is therefore questionable whether these strategies can be financed by the farmers. The costs of S 64, for instance – control of rice leaf folder and stem borer – amount to over 350 Pesos ha^{-1}, making this measure considerably more expensive than the most expensive control strategy used by the farmers.

Therefore, if the use of control thresholds is to be economically feasible, costs must be as low as possible, control effectiveness at least equal to that of the strategies used by the farmers, and maximum returns must be derived from the alternative use of the cash funds saved.

4.3.1.5 Benefit to Be Derived from Infestation Forecasts

The final level of informational improvement would be reached when reliable infestation prognoses are available. Then it would be possible to select the appropriate strategy not only for an infestation situation, but for a specific infestation event. The potential benefit of forecasts can be derived from the average maximum net returns achieved in the individual infestation events and the returns from the initial strategy. This difference in net returns will be computed separately for resistant varieties, since the overall benefit depends on the probability of occurrence of one or the other infestation.

The point of departure for computing the benefit in the case of resistant varieties will be the best control strategy. In this context, the informational value of an infestation prognosis must be regarded as equal to the informational value of economically defined control thresholds (Chap. 3.1). (Actually the control thresholds used in this study do not correspond to the economic definition of control thresholds since they do not necessarily lead to the optimum economic result).

In the case of non-resistant varieties, the maximum net returns over all possible events will serve as a measure of comparison.

For resistant varieties, as shown in Table 56, the average net returns from the respective best strategies amount to 5118 Pesos ha^{-1}. Comparing this with the best strategy S 9 (Table 54), and including the alternative use of the cash funds saved, we obtain an additional net return of 65 Pesos ha^{-1}.

Looking at the individual infestation events, it will be found that in five cases the economically correct decision would be to dispense with control. In three cases, strategy S 7 could be used and in five further cases, strategy S 9. Therefore, in eight cases, both an infestation forecast as well as control thresholds, the economic results of which are equal to those of the respective best measure available to date, would lead to additional net returns. If the control effect can be increased by improving the application technique, then a further increase in

Table 56. Economic parameters of superior strategies, resistant varieties, Nueva Ecija[a]

Economic coefficients	Infestation event[b]													
	1	2	3	4	5	6	7	8	9	10	11	12	13	$\bar{x}$
Net returns ($P\ ha^{-1}$)	4982.2	4921.5	5007	4929	5104.2	5320.7	5326.9	5351.1	5111.0	5082.7	5421.0	5003.0	4973	5118.0
Yield ($kg\ ha^{-1}$)	4219.0	4118.0	4217	4145	4253.0	4378.4	4382.5	4402.2	4169.2	4233.0	4457.4	4195.0	3192	4181.7
Costs of pest control	182.6	182.6	137.0	182.6	137.0	0	0	0	0	182.6	0	137.0	182.6	101.8
Residual loss ($kg\ ha^{-1}$)[c]	409.0	510.0	411.0	452.0	375.0	249.6	246.0	226.0	459.0	315.0	171.0	433.0	636.0	376.0
Potential loss ($kg\ ha^{-1}$)[d]	1178	757.0	1048	935.0	601.0	249.6	246.0	226.0	459.0	510.0	171.0	877.0	786.0	618.0
Strategy no.	S9	S9	S7	S9	S7	S70	S70	S70	S70	S9	S70	S7	S9	–

[a] The opportunity costs of cash have not been taken into account
[b] From Table 49
[c] Potential loss + marginal yield of cash – yield after control
[d] Potential loss + marginal yield of cash – yield in the case of non-control

Table 57. Economic parameters of superior strategies, non-resistant varieties, Nueva Ecija, rainy season[a]

Economic coefficients	Infestation event[b]								
	1	2	3	4	5	6	7	8	$\bar{x}$
Net returns (P ha^{-1})	4769.1	4423.8	4279.0	3629.0	4503.9	5338.8	4991.0	1613.1	4199.1
Yield (kg ha^{-1})	4112.6	4044.2	4029.4	3207.3	4054.8	4424.3	4150.6	1542	3965
Costs of pest control	314.0	314.0	314.0	295.9	314.0	91.3	91.3	179.8	–
Residual loss (kg ha^{-1})[a]	515	584	599	1421	573	204	477	2786	894
Potential loss (kg ha^{-1})[a]	1004	3320	4325	2270	2785	366	651	3802	2315
Strategy	S29	S29	S29	S29	S29	S6	S6	S23	

[a] Refer to Table 56
[b] From Table 49

Table 58. Matrix of infestation events, resistant varieties, Camarines Sur 1976–1982

Pest	Infestation event									
	1	2	3	4	5	6	7	8	9	10
Rice whorl maggot[a]	14.5	14.5	21.7	21.7	0	0	17.44	21.7	6.48	16.68
Army worm, etc.,[a] leaf feeding	0	0	0	0	0	0	10.44	0	1.18	3.10
Army worm, etc.,[a] leaf feeding pests (gen. phase)	0	0	0	0	0	0	20.51	0	2.44	15.46
Rice caseworm[a]	0	0	0	0	0	0	2.97	0	1.59	14.61
Stem borer[b] (gen. phase)	9.09	7.27	0	7.45	3.63	0	0.14	9.09	6.69	9.33
Stem borer[b] (ripening phase)	10.42	10.42	0	0.63	0	0	0	8.33	5.08	9.67
Rice bug[c]	7.46	7.46	7.46	5.97	4.47	7.46	2.17	7.46	0	0
Year	1976	1977	1977	1978	1979	1980	1981	1981	1981	1982
Season	RS	DS	RS	RS	RS	RS	DS	DS	RS	DS

[a] Percent infested leaves
[b] Percent dead hearts/white heads
[c] No. of nymphs and adult bugs (m^{-2})

net returns can even be achieved. Table 56 shows that the potential loss averages almost 650 kg ha^{-1}. The residual loss after control has been carried out still amounts to about 400 kg ha^{-1} so that in each case the control effectiveness of the economically most feasible measure amounts to about 40% only.

If we study the situation in non-resistant varieties, we will arrive at similar results (Table 58). If in each case the best strategy is chosen, the resulting net returns will on the average be 104 Pesos higher than those derived from constant application of strategy S 29.

Looking at the individual data, we will find that in three infestation events, S 29 does not result in the maximum net returns. These events are characterized by the joint occurrence of tungro virus and brown planthopper. The residual crop loss after application of the respective best strategy averages 889 kg ha^{-1} and is thus considerable. Relating this residual loss to the potential loss of 1424 kg ha^{-1} (about 30% of the potential yield), we arrive at an effectiveness of control of 38%, which is almost identical with the situation in resistant varieties.

It is therefore clear in the case of both infestation situations that with control effectiveness remaining unchanged, the benefit to be derived from improving the current control thresholds would amount to only 80 to 100 Pesos ha^{-1} and would thus be small. On the other hand, if control effectiveness can be increased to perhaps 60%, additional returns of the order of 200–300 Pesos ha^{-1} could be realized.

4.3.2 Results Computed for the Camarines Sur Area

4.3.2.1 Model Assumptions

The results for Camarines Sur will be presented in the same way as the results for Nueva Ecija in the preceding chapter.

As opposed to Nueva Ecija and Iloilo, it was not always possible here to rely on infestation data from untreated plots. Instead, data from an observation station of the official crop protection warning service had to be used. Since the data concerned were obtained from treated farmers' fields, they were converted into infestation data of untreated plots (i.e. non-control) on the basis of the avarage control effectiveness of the farmers' strategies.

As shown in Table 58, the infestation events are characterized to a marked degree by the occurrence of the rice whorl maggot during the early growth stage and of stem borer and rice bug at a later stage. In four infestation events, stem borer and rice bug both both reached their control threshold (5% dead hearts and 4 bugs m^{-2}). In three further cases, the latter pest reaches its control threshold alone and stem borer and army worm (generative growth stage) reach their control threshold in one case each. In one further event (1982), the two latter pests both reached their control thresholds.

In the case of non-resistant varieties, there is a significant difference between rainy and dry seasons in the level of attack by the brown planthopper (cf. Chap. 2.2.5.1.1, Table 21b). For the tungro virus, a significant difference of this kind was not found so that it was possible to use the averages of the individual years.

It can be seen in Table 59 that during the period from 1977–1980, the brown planthopper population became extremely abundant during the rainy season whilst the other years were characterized only by a medium level of tungro incidence. The economic parameters of the Camarines Sur area are summarized in Table 60. The data have been obtained from the same sources as those used for the Nueva Ecija area. The computation of marginal yields is based on the pro-

Table 59. Matrix of infestation events, non-resistant varieties, Camarines Sur, rainy and dry seasons, 1972–1982

Pest	Infestation event							
	1	2	3	4	5	6	7	8
Rainy season								
Tungro virus[a,b]	43.0	10.95	17.3	47.17	14.33	44.81	36.66	55.24
Brown plant-hopper[c]	0	0	0	75.16	337.5	130.0	203.0	0
Dry season								
Brown plant-hopper[c]	0	0	0	0	34.16	90.0	42.0	0
Year	1972	1973	1975	1977	1978	1979	1980	1981

[a] Percent infested hills
[b] Virus infection in the dry season is identical
[c] No. per hill
Source: BRES – Annual Reports, Entomology Division

Table 60. Economic parameters used in the calculations for Camarines Sur

	Crop price (P kg^{-1})	% Interest per production period	Marginal yield of cash in % of yield per Peso	Potential yield (kg ha^{-1})	Financial limit (P ha^{-1})
Rainy season	1.0	11.0	0.136	4460	205 (120)
Dry season	0.95	11.0	0.126	4815	205 (120)

Source: cf. Chapter 4.2.4

duction function for herbicides listed in Table 48 (Sect. 4.2.4.8). According to this function, every additional unit of money (Pesos) expended on herbicides results in a yield increase of 6.1 kg. If we relate this yield to the potential rainy reason yield (4460 kg ha^{-1}) and the corresponding dry season yield (4815 kg ha^{-1}), marginal yields of 0.136 and 0.126% are arrived at. Since it cannot be assumed that marginal yields of this magnitude are achieved in the case of all cash funds saved, the maximum sum expended on chemical weed control (120 Pesos ha^{-1}) was taken as the upper limit.

4.3.2.2 The Optimum Strategy

The complete data matrix for Camarines Sur covers a total of 1260 data results from ten infestation events in resistant varieties, eight infestation events in non-

Table 61. Basic matrix for computing the benefits of warning notices in Camarines Sur, rainy and dry season[a]

Strategies	Events			
	Dry season		Rainy season	
	E_1	E_2	E_1	E_2
$SOPT_{nres}$: S36	3539	2981	3638	3419
S36	(3539)	(2981)	(3637)	(3419)
$SOPT_{res}$: S9	3922	1835	4027	2465
S70	(4456)	(2307)	(4506)	(3046)
$SOPT_{res}^{nres}$: S15	3749	2882	3849	3363
S15	(3935)	(2959)	(4023)	(3515)
$SFARM_{res}^{nres}$: S6	3760	2554	3859	3229
S6	(4187)	(2820)	(4258)	(3556)

[a] Figures in brackets refer to the alternative strategy taking into account opportunity costs of cash

Source: Author's computations

resistant varieties and the 70 strategies available for selection. Table 61 is a section of this data matrix.

It will be seen from the data that, taking into account the opportunity costs of cash, the highest average net return (according to the definition given above) of 3503 Pesos ha^{-1} (rainy season) and 3907 Pesos ha^{-1} (dry season) respectively is achieved by means of strategy S 6 (two applications of Azinophosethyl at a reduced rate during the early growth stage) (Table 62). These figures must be viewed with certain reservations in as much as the economic result of S 6 is greatly affected by the comparatively high marginal yield (14%) of the cash funds saved, since the costs of this measure amount to only slightly more than 90 Pesos ha^{-1}. For this reason, the figures not including the opportunity costs of cash are more transparent with respect to the effectiveness of the crop protection measure. Here we find that, as in the case of the Nueva Ecija results, strategy S 15 amounting to 3285 Pesos ha^{-1} (rainy season) and 3606 Pesos ha^{-1} (dry season) respectively must be regarded as the optimum strategy. If Carbofuran is worked into the soil before transplanting (S 15), then under dry-season conditions, this strategy meets the requirement of a certain minimum yield being achieved and at the same time conforms to the set liquidity constraints. The yield obtained in the most unfavourable case is still more than 250 kg higher than the computed minimum yield of 1969 kg ha^{-1} (Table 62). On the other hand, none of the available strategies achieve the minimum rainy season yield of 2551 kg ha^{-1} although S 45 with a minimum yield of almost 2400 kg comes closest to meeting this requirement. Therefore there is no strategy which will conform to both the yield and the liquidity constraints during the rainy season. As in the case of the Nueva Ecija investigation area, strategy S 15 seems to result in the greatest economic advantages under the current conditions.

On the other hand, if the decision-maker goes by the maximin rule, strategy S 36 with a minimum net return of 1800 Pesos ha^{-1} net of crop protection costs

Table 62. Average net returns based on alternative crop protection strategies in Pesos ha^{-1}, Camarines Sur, rainy and dry season

Strategies	Costs of pest control (P ha^{-1})	Rainy season					Dry season				
		With opportunity cost of cash			Without opportunity cost of cash		With opportunity cost of cash			Without opportunity cost of cash	
		Net returns (P ha^{-1})	Min. yield[a] (kg ha^{-1})	Marg. yield (%)	Net returns (P ha^{-1})	Min. yield[a] (kg ha^{-1})	Net returns (P ha^{-1})	Min. yield[a] (kg ha^{-1})	Marg. yield (%)	Net returns (P ha^{-1})	Min. yield[a] (kg ha^{-1})
S 5	218.0	3043.5	1331.1	0	3043.5	1331.1	3430.9	1657.4	0	3430.9	1657.4
S 6	89.1	3503.3	1517.8	14.03	3157.3	1331.1	3907.1	1872.8	13.0	3544.3	1657.4
S 7	133.6	3064.3	580.6	8.64	2874.3	545.7	3441.0	874.3	8.01	3240.8	809.5
S 8	204.6	2832.1	545.9	0.04	2831.1	545.7	3199.4	809.8	0.04	3198.4	809.5
S 9	178.1	2951.8	563.4	3.25	2878.6	545.7	3323.4	833.8	3.01	3246.4	809.5
S10	232.5	3030.6	1331.1	0	3030.6	1331.1	3418.9	1657.4	0	3418.0	1657.4
S12	154.3	3277.4	1412.8	6.14	3123.7	1331.1	3672.4	865.4	5.69	3511.6	809.5
S13	231.4	3057.8	1331.1	0	3057.8	1331.1	3445.9	1657.4	0	3445.9	1657.4
S15	155.4	3446.8	1948.9	6	3285.6	1838.6	3769.0	2227.8	5.56	3605.8	2110.5
S16	305.3	3215.5	2075.0	0	3215.5	2075.0	3504.3	2321.7	0	3504.3	2321.7
S17	155.4	3299.2	1713.3	6	3147.3	1616.3	3617.5	1974.6	5.56	3463.2	1870.6
S23	175.4	3357.8	1989.5	3.59	3261.7	1920.6	3644.0	2233.5	3.32	3547.9	2161.7
S29	306.2	3257.8	2272.3	0	3257.8	2272.8	3558.6	2547.9	0	3558.6	2547.9
S30	242.0	3198.2	1997.4	0	3198.2	1997.4	3493.2	2248.9	0	3493.2	2248.9
S32	441.8	3140.4	2345.3	0	3140.3	2345.3	3423.1	2602.4	0	3423.1	2602.4
S35	638.3	2923.1	2270.7	0	2923.1	2270.7	3189.8	2535.1	0	3189.8	2535.1
S36	288.6	3260.1	2251.2	0	3260.1	2251.2	3528.0	2514.1	0	3528.0	2514.1
S45	612.5	3114.2	2395.2	0	3114.2	2395.2	3387.7	2666.9	0	3378.7	2666.9
No control	0	3171.1	624.9	14.52	2873.8	545.7	3552.6	918.5	13.46	3236.9	809.5

[a] Min. yield = lowest yield realized

would have to be selected during the rainy season, and strategy S 29 with 1932 Pesos ha^{-1} during the dry season. Both strategies would exceed the set liquidity restrictions by almost 30%. Unless pesticide prices were state subsidized, only a small number of farms would therefore be able to afford them.

Finally, comparing the net returns from crop protection measures with a situation without control, we will find that in this case, too, the results are greatly influenced by the opportunity costs of cash. If these costs are taken into account, the marginal net returns, if control measures are instituted, amount to 322 Pesos ha^{-1} in the rainy season and to 355 Pesos ha^{-1} in the dry season. Neglecting the opportunity costs of cash, the marginal net return from control will amount to 412 Pesos ha^{-1} in the rainy season and to 369 Pesos ha^{-1} in the dry season. In Camarines Sur, chemical crop protection is consequently marginally more profitable in the rainy season than in the dry season.

This superiority is due to the clearly higher incidence of the brown planthopper during the rainy season. Overall, studying the results, it will be noted that the strategies listed in Table 62 differ only marginally with respect to their returns, while there are considerable differences with respect to costs and minimum yields.

The choice of strategy is therefore determined primarily by whether or not the set restrictions are conformed to.

4.3.2.2.1 Comparison Between Optimum Strategy and Farmers' Strategies. If the opportunity costs of cash are taken into account, then, as explained above, strategy S 6, the farmers' strategy involving the lowest costs, will result in the maximum net returns. A comparison between the optimum strategy and the farmers' strategies is therefore necessary only if the opportunity costs are neglected. In this case, the average difference in returns between S 15, the optimum strategy, and S 6 will amount to 128 Pesos ha^{-1} in the rainy season and to only 51 Pesos ha^{-1} in the dry season. With respect to minimum yield, the difference amounts to 635 kg ha^{-1} in the rainy season and to 453 kg ha^{-1} in the dry season. This corresponds to 90% and 50% respectively of the farm household consumption of rice listed in Table 45 (Sect. 4.2.4.6).

If we take into account the uncertainty of the assumption that strategy S 6 will correspond to effective tungro control, then S 9 would be the next best measure. The difference in returns between S 9 and S 15 would amount to 407 Pesos ha^{-1} in the rainy season and to 401 Pesos ha^{-1} in the dry season. As regards the minimum yields achieved by strategy S 9, these amount to 546 kg ha^{-1} in the rainy period and to 810 kg ha^{-1} in the dry period and are thus clearly below the set restrictions of 2551 and 1961 kg ha^{-1} respectively.

Furthermore, a comparison between S 9 and a situation without control shows that the incremental net returns amount to as little as 5 and 10 Pesos ha^{-1} respectively. If we were to assume that the farmers' strategies fail to have any effect in the case of varietal resistance breaking down, then dispensing with a control measure would hardly influence the net return. Since the control methods used by the farmers are not an adequate guarantee in the case of extreme pest density, a basic change of crop protection strategy would be advantageous in Camarines Sur.

4.3.2.3 Benefit of the Warning Service

The following computation of the benefit of the warning service is based on the method described in Section 4.3.1.3. It is consequently necessary to list the infestation situations occurring in resistant and non-resistant varieties. Table 62 shows the optimum strategy in the case of occurrence of the possible events in resistant varieties (E_1) $SOPT_{res}$ and of the possible events in non-resistant varieties (E_2) $SOPT_{nres}$ as well as the optimum strategy overall, $SOPT_{res}^{nres}$ and the best farmers' strategy governing both possible types of events $SFARM_{res}^{nres}$. On the basis of these data, the difference in net returns between the measures recommended by the warning service and any other strategy can be computed.

If E_1 occurs, S 9 would be the appropriate measure to take (Table 62). Compared with S 15, an incremental benefit of 173 Pesos ha^{-1} (rainy season) and 178 Pesos ha^{-1} (dry season) can be realized. If S 6 is assumed to be the initial strategy, the differences amount to 162 Pesos ha^{-1} and 168 Pesos ha^{-1} respectively. In the case of resistant varieties, the best farmers' strategy is therefore marginally more favourable than S 15.

In the reverse case, i.e. in the case of a warning indicating the occurrence of E_2, S 36 will result in the highest average net returns, the difference between this strategy and S 15 or S 6 amounting to 99 or 427 Pesos ha^{-1} respectively, in the rainy season. However, in the dry season, the corresponding differences amount to only 57 or 190 Pesos ha^{-1} respectively. The distinct differences between rainy and dry seasons are attributable to the higher incidence of the brown planthopper during the rainy season. In the case of this pest, the control effectiveness of the farmers' strategies is equal to zero whilst the effectiveness of S 36 was computed as 76% (cf. Chapter 4.2.2.3).

The above computations apply only on the assumption that the warnings always prove correct. As regards the minimum necessary degree of reliability of a warning, on the other hand, this can be computed as shown in Section 4.3.1.3. For this purpose, we will again make use of the graph. Figure 14a shows that up to a probability of 0.14 of varietal resistance breaking down (E_2), strategy S 9 would be economically feasible. This corresponds to a minimum probability of 0.86 of the forecasted durability of varietal resistance (E_1) actually proving correct. If the probability that E_2 takes place is greater than 0.65, use of strategy S 36 is economically justified. For the in-between range, strategy S 15 is superior to the other strategies. It is therefore the strategy involving the least informational risk.

In the dry season, the picture is similar (Fig. 14b). Here, use of S 9 presupposes a reliability of not quite 0.9 and use of S 36 a minimum reliability of 0.45 if it replaces S 6 and of not quite 0.2 if it replaces S 15. Again, strategy S 15 proves to be the crop protection measure involving the least risk.

4.3.2.4 Benefit of Control Thresholds

In view of the high opportunity costs of cash (6 kg of rice for each additional unit of money expended on weed control), the decision to dispense with control and instead put the cash funds saved to an alternative use seems plausible from an

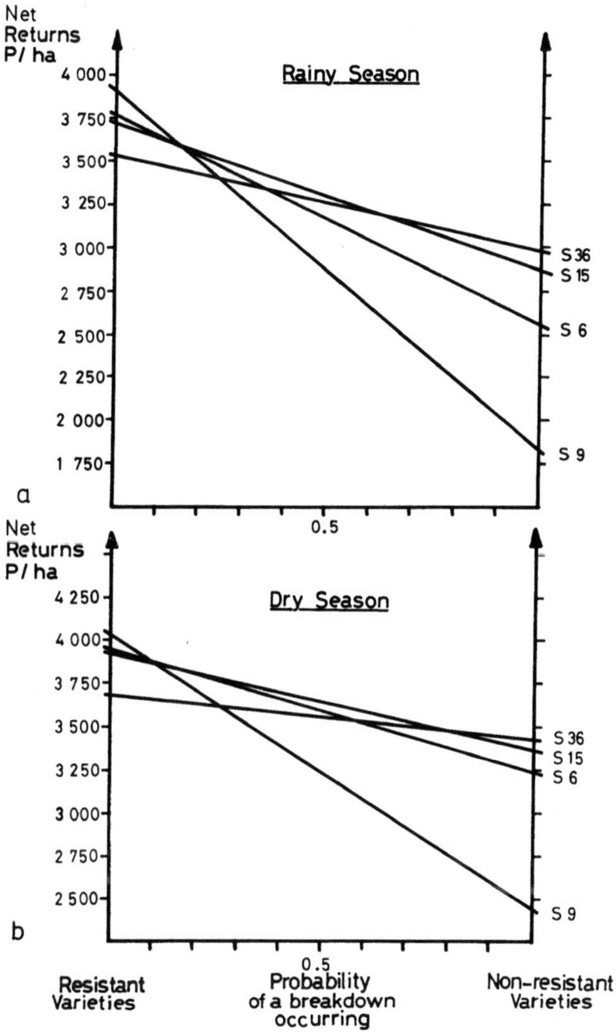

Fig. 14a, b. The optimum strategy as related to the probability of varietal resistance breaking down, Camarines Sur, rainy and dry season

economic point of view. Due to the nature of the infestation situations, the control threshold is reached in every case. Owing to the slight effectiveness of control, use of the control threshold concept leads to economically poorer results. Considering all infestation events, the average net return computed for the control threshold strategy is 254 Pesos ha^{-1} lower than in the case of non-control (Table 63). The best measure used in the field (S 6) results in a return exceeding the control threshold concept on an average of 284 Pesos ha^{-1} and exceeding even non-control by as much as 30 Pesos ha^{-1} in the event that opportunity costs of cash do not exist.

If it is assumed that the cash input does not give rise to any opportunity costs, the control threshold concept will be superior to non-control by 35 Pesos ha^{-1}. Control thresholds, on the other hand, are inferior to the best farmers' strategy by 152 Pesos ha^{-1}.

Table 63. Marginal net returns based on the control threshold concept and farmer's strategies in Pesos/ha resistant varieties, Camarines Sur

Strategy	Infestation event[a]										
	1	2	3	4	5	6	7	8	9	10	$\bar{x}$
With opportunity costs of cash											
Best farmer strategy (S6)	+31.3	+12.0	−103.8	−23.7	−67.9	+242.2	−105.2	+24.3	+24.5	+264.9	+29.9
Control thresholds	−312.4 (S68)	−347.3 (S68)	−170.2 (S63)	−468.6 (S68)	−223.3 (S63)	+90.0 (S60)	−184.4 (S63)	−296.6 (S68)	−630.9 (S62)	+12.0 (S56)	−253.7
No control	4205.2	4253.1	4275.9	4140.6	4670.7	3670.7	4629.4	3915.8	4586.2	3222.0	4156.9
Without opportunity costs of cash											
Best farmer strategy (S9)	+206.7	+180.8	−22.3	+42.8	−41.0	+501.0	−22.2	+178.4	+37.1	+820.8	+187.7
Control thresholds	+116.0 (S68)	+87.2 (S68)	−39.8 (S63)	−48.4 (S68)	−73.3 (S63)	+94.1 (S60)	−39.8 (S63)	+93.1 (S68)	−154.2 (S62)	+315.7 (S56)	+35.1
No control	3776.8	3818.6	3838.5	3720.4	4183.3	3309.5	4147.2	3524.1	4109.5	2918.3	3734.6

[a] See Table 58

Source: Author's computations

Looking at the individual results, it will be noted that the strategies used in the field are superior to the control threshold concept in every case. In three infestation events (Nos. 3, 5, 7) – here the rice bug exceeds its control threshold – even strategy S 9 would be less favourable than non-control. This is due to the low effectiveness of the farmers' strategies in controlling pests which occur in later growth stages. It is only in 50% of the infestation events that the control threshold concept proves superior to non-control. In the case of the rice bug (S 63), the stem borer (S 62) and stem borer and rice bug (one case only, S 68), the cost of control based on control thresholds could not be met.

To sum up, we can therefore say that in infestation events in which the same pest often reaches its threshold, control thresholds at the present level of effectiveness are less economical than prophylactic measures.

4.3.2.5 Benefit of Infestation Forecasts

In our examination of the potential benefit of infestation forecasts and reliable control thresholds, opportunity costs of cash have not been included. By including these costs, the results of the crop protection measures would largely be influenced by the high value of the alternative use of cash. Table 64 gives the data of the best strategy for each infestation event in resistant varieties. The average return amounts to 3943 Pesos ha^{-1} and is thus only 20 Pesos higher than that yielded by the constant use of S 9. In four cases it is not exceeded by any other strategy, and in two further cases "non-control" proves superior to all other existing alternatives.

The average residual loss of almost 350 kg ha^{-1} shows that potentially an improvement of the profitability of crop protection would still be possible. On the basis of an average potential loss of 761 kg ha^{-1}, the average control effectiveness of all the respective best measures is computed as 54%, and must therefore be regarded as unsatisfactory.

On the other hand, all the strategies listed in Table 64 conform to both the liquidity restrictions and to the set minimum yield restrictions. An improvement in effective control involving higher costs cannot be undertaken unless the resulting incremental returns exceed the opportunity costs of the additional cash needed.

In the case of the events in non-resistant varieties, there is again only a marginal increase in net returns ranging from 2981 Pesos ha^{-1} (Table 62) to 3015 Pesos ha^{-1} (Table 65). These figures refer to the rainy season; in the dry season, due to the lower incidence of the brown planthopper, the net returns are even lower. On an average, taking into account all infestation events in non-resistant varieties, S 36 is superior to all other strategies in five out of eight cases, as shown in Table 65. In the case of one event, the correct strategy would be S 6, the cheap method used in thc field; in the case of one further event, the appropriate strategy would consist in the use of Carbofuran before transplanting, and in the case of event No. 8, S 29 would be the appropriate strategy to chose.

Here again, the ratio between the average residual loss of over 1000 kg and the potential loss of almost 2500 kg ha^{-1} shows that an improvement in control effectiveness would have economic advantages. Such an improvement should

Table 64. Economic parameters of superior control strategies, resistant varieties, Camarines Sur, rainy and dry seasons[a]

Economic coefficients	Infestation events[b]										
	1	2	3	4	5	6	7	8	9	10	$\bar{x}$
Returns (P ha^{-1})	3978.5	4004.8	3876.1	3829.4	4183.3	3810.5	4147.2	3702.1	4162.1	3739.1	3943.4
Yield (kg ha^{-1})	4203.6	4153.8	4055.4	4017.4	4183.3	4072.6	4147.2	3924.7	4318.7	4040.7	4111.7
Costs of pest control	178.1	112.1	155.4	155.4	0	178.1	0	178.1	133.6	178.1	–
Residual loss (kg ha^{-1})[a]	256.0	306.0	405.0	443.0	277.0	387.0	313.0	535.0	141.0	419.0	348.0
Potential loss (kg ha^{-1})[a]	683.0	641.0	622.0	740.0	277.0	1151.0	313.0	936.0	351.0	1542.0	761.0
Strategy	S9	S27	S15	S15	S70	S9	S70	S9	S7	S9	–

[a] Cf. Table 56
[b] See Table 58
Source: Author's computations

Table 65. Economic parameters of superior control strategies, non-resistant varieties, Camarines Sur, rainy and dry season[a]

Economic coefficients	Infestation events[b]								
	1	2	3	4	5	6	7	8	$\bar{x}$
Returns ($P\ ha^{-1}$)	3756.6	4146.4	4043.6	1788.2	2785.6	1800.4	2034.9	3764.6	3015.0
Yield ($kg\ ha^{-1}$)	4204.7	4280.3	4266.7	2251.7	3261.5	2279.8	2520.8	4286.4	3418.9
Costs of pest control ($P\ ha^{-1}$)	288.6	89.1	155.4	288.6	288.6	288.6	288.6	306.2	–
Residual loss[a] ($kg\ ha^{-1}$)	255.0	180.0	193.0	2209.0	1199.0	2180.0	1939.0	174.0	1041
Potential loss[a]	1706.0	587.0	808.0	3795.0	2901.0	3914.0	3733.0	2133.0	2447
Strategy[a]	S36	S6	S15	S36	S36	S36	S36	S29	–

[a] Cf. Table 56
[b] See Table 59
Source: Author's computations

not, however, increase the costs of these measures to any marked degree. In six out of eight cases, costs already exceed the set liquidity restrictions so that it would only be possible to realize the measures if pesticides are subsidized by the government. In the two instances where it is not the case, an improvement in control effectiveness would be an advantage.

Overall, the results show that the benefit of infestation prognoses, both in resistant and non-resistant varieties, would be only marginal without a simultaneous improvement in the technical and economic efficiency of pest control methods.

4.3.3 Results Computed for the Iloilo Area

4.3.3.1 Model Assumptions

In the Province of Iloilo, a total of 15 infestation events on untreated plots were available for our computations. Details are given in Table 66. It will be noted that composition and intensity of infestation vary greatly.

The control threshold was most frequently reached by rice bug, together with the rice leaf folder. The latter definitely exceeded its threshold (25% infested leaves) in one further case, reaching an infestation intensity of 48%. Overall, it is in only 7 out of 15 cases that the control threshold is surpassed.

In non-resistant varieties, the situation is different than that in the other two cultivation areas. On the untreated plots of the insecticide trials in which the susceptible IR-22 variety was used, hardly any tungro infestation occurred. However, during the 1980 rainy reason, the population of the brown planthopper was comparatively high, amounting to 38 insects per hill (Table 67).

Table 66. Matrix of infestation events in resistant varieties, Iloilo 1976–1982

	Infestation events														
	1	2	3	4	5	6	7	8	9	10	11	12	13	14	15
Rice whorl maggot[a]															4.06
Army worm, etc., leaf feeding pests[a] (veg. phase)													2.54	2.06	1.5
Rice caseworm	21.0				54.0						48.0	8.25	1.19	0.11	
Army worm, etc., leaf feeding pests (gen. phase)														4.15	3.23
Rice leaf folder[a] (gen. phase)		13.0						13.0			11.0	4.95	2.11	2.67	1.55
Stem borer[b] (gen. phase)							4.6						0.02	1.31	5.49
Stem borer[b] (ripening phase)	7.4	3.4	2.7		4.4		2.1	5.6	4.8	2.4		2.54	0.04	0.35	2.33
Rice bug[c]	4.7			14.0	4.1	8.0	5.0		5.5	1.6					
Season	RS	RS	DS	DS	RS	RS	RS	DS	DS	RS	RS	RS	DS	RS	RS
Year	1976	1977	1978	1978	1978	1978	1978	1979	1979	1979	1979	1980	1981	1981	1982
Source	1979 A (63)	1979 A (55)	1979 A (64)	1979 A (65)	1979 A (58)	1979 A (66)	1979 A (56)	1979 A (59)	1979 A (67)	1980 A (66)	1980 A (70)	B	B	B	B

[a] Percent infested leaves
[b] Percent dead hearts/white heads
[c] No. of bugs (m^{-2})
A = IRRI Insecticide Evaluation Report, figures in brackets = nos. of tables
B = Own trials

Table 67. Matrix of infestation events in non-resistant varieties, Iloilo 1975–1980

	Infestation event					
	1	2	3	4	5	6
Rainy season						
Tungro virus[a]	4.0	9.4	2.0	1.1	0.0	0.0
Brown planthopper[b]	11.0	14.5	3.7	8.2	10.1	38.3
Dry season						
Tungro virus[a]	4.0	9.4	2.0	1.1	0.0	–
Brown planthopper[b]	0.0	3.0	1.1	2.3	0.4	–
Year	1975	1976	1977	1978	1979	1980

[a] Percent infested hills
[b] No. per hill
Source: VRES – Annual Reports, Entomology Division

Table 68. Economic parameters of model computations for Iloilo

	Crop price ($P\ kg^{-1}$)	Interest per production period (%)	Marginal yield of cash in % of yield per Peso	Potential yield ($kg\ ha^{-1}$)	Financial limit ($P\ ha^{-1}$)
Rainy season	0.93	15.0	0.057	4584	230
Dry season	0.95	15.0	0.08	3250	230

Source: cf. Chapter 4.2.4

According to the official crop protection service, outbreaks of the brown planthopper also occurred in the Province of Iloilo (Bayoneta 1981, pers. comm.), but figures on this infestation event were not available.

The economic parameters for our model computations are summarized in Table 68. Again, attention should be drawn to the marginal yield, amounting to 12.14 kg, to be obtained by an additional expenditure on N fertilizer. As in our computation for Nueva Ecija, a price of 4.58 Pesos kg^{-1} has been assumed for *N* fertilizer, so that for Iloilo we arrive at an incremental yield of 2.65 kg for every additional unit of money spent. Relating this marginal yield to the potential yield of 4584 (rainy season) and 3250 kg ha^{-1} (dry season), we obtain the values listed in Table 68.

The other parameters were obtained from the same source as the data for the other two areas.

4.3.3.2 Optimum Strategy and Comparison with Farmers' Measures

Studying the infestation situations, it is clear that in Iloilo the optimum strategy is influenced considerably by the events in resistant varieties.

Table 69. Average net returns from alternative crop protection strategies in Pesos ha^{-1}, Iloilo, rainy and dry season

Strategies	Costs of pest control (P ha^{-1})	Rainy season					Dry season				
		With opportunity costs of cash			Without opportunity costs of cash		With opportunity costs of cash			Without opportunity costs of cash	
		Net returns (P ha^{-1})	Min. yield[a] (kg ha^{-1})	Marg. yield (%)	Net returns (P ha^{-1})	Min. yield[a] (kg ha^{-1})	Net returns (P ha^{-1})	Min. yield[a] (kg ha^{-1})	Marg. yield (%)	Net returns (P ha^{-1})	Min. yield[a] (kg ha^{-1})
S 1	129.7	3768.3	3153.7	4.86	3688.9	3007.6	2762.5	2465.4	6.82	2671.5	2308.5
S 2	194.6	3630.2	3059.0	1.71	3602.8	3007.6	2622.9	2363.6	2.41	2591.3	2308.5
S 3	126.9	3751.7	3157.7	4.99	3670.8	3007.6	2751.6	2470.3	7.01	2659.2	2308.5
S 4	190.4	3711.3	3065.3	1.92	3637.9	3007.6	2654.1	2370.5	2.69	2617.9	2308.5
S 5	225.9	3681.3	3329.2	0.2	3678.6	3328.5	2635.2	2542.8	0.27	2634.2	2536.0
S 6	92.3	3934.8	3550.5	6.67	3798.4	3328.5	2884.7	2773.6	9.37	2754.8	2536.0
S 7	138.4	3767.8	3141.1	4.44	3724.4	3007.6	2780.1	2452.3	6.23	2694.8	2308.5
S 9	184.6	3782.4	3073.7	2.2	3744.2	3007.6	2784.2	2379.8	3.09	2696.4	2308.5
S15	161.0	3858.5	3543.1	3.57	3790.6	3421.0	2786.3	2663.9	4.69	2769.4	2544.6
S17	161.0	3785.3	3527.0	3.57	3715.1	3405.5	2726.0	2661.9	4.69	2709.4	2542.3
S23	181.7	3753.0	3545.0	2.34	3710.0	3464.0	2692.5	2634.4	3.28	2647.7	2550.8
No control	0.0	3942.0	3342.6	11.14	3764.9	3007.6	2964.4	2669.5	15.64	2762.4	2308.5

[a] Min. yield = lowest yield realized

Table 69 again gives a section of the complete data matrix. If the opportunity costs of cash are taken into account, in the case of non-control, the alternative use of the cash funds would result in a yield increase of 11% (rainy season) and 15% (dry season). These yield increases are higher than the average losses due to pests. In this case, non-control would therefore lead to the most favourable economic results. If we set the opportunity costs equal to zero, then strategy S 15 resulting in average net return of 2769 Pesos ha^{-1} is the superior strategy in the dry season. The spray strategy yielding the maximum net returns is S 6. But since the strategy concerned is not typical for the Iloilo area, it has been replaced by the next best strategy for the purpose of further computation.

Of all the measures used by the farmers in Iloilo (S1–S5), strategy S1 yields the best results, even though the average net return achieved is still 100 Pesos ha^{-1} lower than in the case of S15. Since all the farmers' strategies conform to the set restrictions of 1932 kg ha^{-1} in the rainy season, and 1896 kg ha^{-1} in the dry season, it would not appear advisable to replace the farmers' strategies with S15. Rather, it will be necessary to identify the cases in which control would be expedient and the control measures to be used in these cases.

As regards the warning service, the risk involved in incorrect information is considerably smaller in Iloilo so that here activities should focus on preventing unnecessary crop protection measures and promoting the alternative use of the amount of cash saved.

4.3.3.3 Benefit of the Warning Service

Studying the average net returns and minimum yields (Table 70), it becomes obvious that the economic differences between the various strategies are comparatively small. Furthermore, no clear difference between crop losses in resistant and non-resistant varieties will be seen. For this reason, warnings cannot really bring about any great additional benefit with respect to the optimum strategy in each case. However, the additional benefit with respect to the best farmer's strategy in the Iloilo area (S1) is higher, since the latter is inferior to the optimum strategy in any case (S 6).

Table 70 again gives the optimum strategy for each of the two types of event. In the case of events E_1 occurring (resistant varieties), the correct solution would be strategy 9, if E_2 occurs, it would be strategy 6. If E_1 occurs, then compared with S1, the best farmers' strategy, an average incremental return of 166 Pesos ha^{-1} can be achieved in the rainy season. If E_2 occurs, we arrive at a difference of 212 Pesos ha^{-1}. However, in both cases, no account has been taken of the opportunity costs of cash. Including the opportunity costs of cash, the resulting figures will be 125 and 255 Pesos ha^{-1}. In the dry season, the corresponding difference in net returns amounts to 104 and 222 Pesos ha^{-1}, excluding the opportunity costs of cash, and to 158 and 245 Pesos ha^{-1}, including the opportunity costs.

Since the best farmers' strategy is still clearly inferior to the optimum strategy governing both types of events, the benefit of the warning service remains positive.

On the other hand, if the optimum strategy overall is used as a basis of comparison for the various alternatives, the difference in net returns becomes clearly smaller. This fact is important in as much as the costs of the overall optimum

Table 70. Basic matrix for computing the benefits of warning notices in Iloilo, rainy and dry seasons[a]

Strategies	Events			
	Rainy season		Dry season	
	E_1	E_2	E_1	E_2
$SOPT_{nres}$: S6	3774	3822	2708	2801
S6	(3896)	(3947)	(2944)	(2985)
$SOPT_{res}^{b}$: S9	3933	3556	2797	2524
S56	(3982)	(3926)	(2944)	(2985)
$SFARM_{res}^{nres^c}$: S1	3767	3610	2693	2579
S1	(3857)	(3692)	(2786)	(2740)

[a] Figures in brackets refer to the alternative in which the opportunity costs of cash are taken into account
[b] $SOPT_{res}$ also corresponds to the optimum strategy governing both types of events
[c] In the dry season, the figures in brackets refer to S3

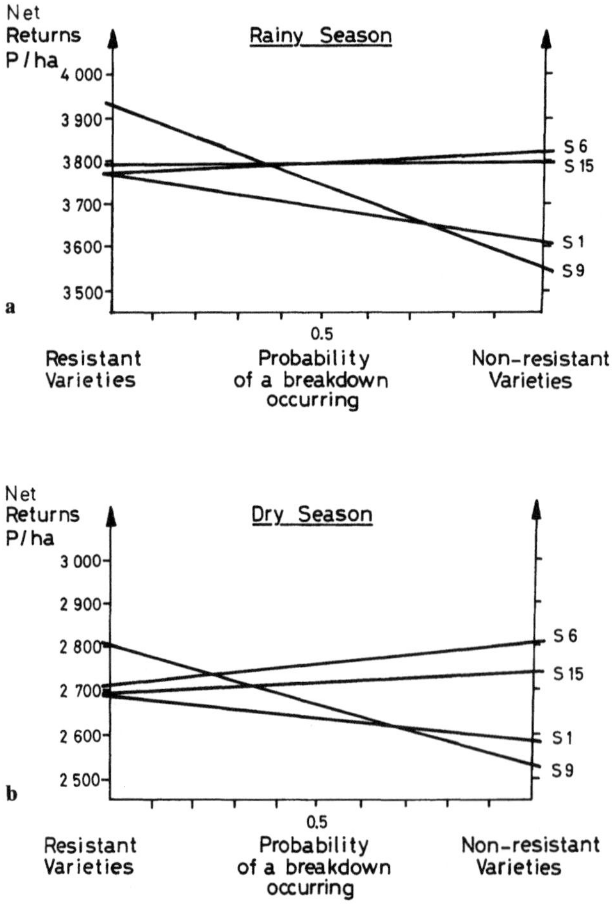

Fig. 15a, b. The optimum strategy as related to the probability of varietal resistance breaking down, Iloilo, rainy and dry season

strategies are lower than those of the best farmers' strategy in the Iloilo area (S1). To improve the situation, it would suffice to change the farmers' strategies. For this reason, the warning service programme will only yield a marginal net return, and its benefit must therefore be seen in the long-term improvement of the decision basis rather than in the provision of current decision making support.

As Table 70 shows, there is no great difference between events E_1 and E_2 in the average net return from the strategies. Incorrect information would consequently result in only marginal economic disadvantages. Figure 15 shows that even if there is a 75% chance of E_2 occurring, S 9 would still be inferior to S 1 in economic respects (dry season). If the strategy recommended by the warning service is based on E_2, S 9 will be superior to S 1 in every case. In Fig. 15a and b, the connecting line for S 6 remains above S 1 throughout the entire range of probabilities.

Regarding the above computations for the Iloilo investigation area, attention must once more be drawn to the low pest incidence in resistant varieties. However, it cannot be entirely ruled out that mass outbreaks of the brown planthopper or a tungro virus infection will also occur in this area in future. In view of the risk aversion of government decision-makers, a surveillance programme can therefore also produce short-term benefits.

4.3.3.4 Benefit of Control Thresholds

The control effectiveness of chemical measures being low when one pest reaches its control threshold, whilst comparatively high costs are incurred when two pests reach their damage threshold, means that in Iloilo as well, the control threshold concept is inferior to the other prophylactic measures. This inferiority becomes all the more pronounced, the higher the mean crop loss, the smaller the variance of loss, and the greater the uniformity of the infestation structure through time. On the other hand, the inferiority of the control threshold concept as compared to non-control will be all the greater, the lower the mean crop loss and the higher the yields resulting from the alternative use of cash. Considering all infestation events and taking into account the opportunity costs, non-control will on an average result in an increase in net returns of 98.4 Pesos ha^{-1} (Table 71). If we exclude the opportunity costs, the difference will amount to only 36 Pesos ha^{-1}.

As compared with the best farmers' strategy in Iloilo (S1), application of the control threshold concept will increase the average net returns by almost 27 Pesos. Excluding the opportunity costs, the net returns will be 12 Pesos ha^{-1} lower. However, if we take into account farmers' strategies from all investigation areas, then S 9 will be superior to the control threshold concept by 95 Pesos ha^{-1}, excluding opportunity costs respectively.

Strategy S 9 involves one further pesticide application during the final growth stage so that better control of the rice bug is thus achieved than by means of the Iloilo strategies.

In view of the increasing heterogeneity of infestation structure, the benefit of the control threshold concept could definitely be increased if more realistic threshold values were developed.

Table 71. Marginal net returns based on the control threshold concept and farmer strategies, Pesos

Strategies	Infestation events[a]					
	1	2	3	4	5	6
Without opportunity costs of cash						
Farmer strategy (Iloilo S 1)	−43.2	−51.5	−242.6	−227.4	+153.4	−244.8
Best farmer strat. (overall S 9)	+276.1	+93.4	−288.8	−44.5	+638.7	−188.8
Control threshold	−160.7 (S 63)	0.0 (S 70)	0.0 (S 70)	−71.0 (S 63)	−75.3 (S 55)	−153.7 (S 63)
No control	3607.5	3917.6	4386.8	3884.6	3170.0	4151.9
With opportunity costs of cash						
Farmer strategy (Iloilo S 1)	+35.1	+43.8	−122.2	−124.7	+468.1	−126.8
Best farmer strat. (overall S 9)	+386.2	+232.1	−103.9	+95.1	+705.6	−23.7
Control threshold	−77.7 (S 63)	0.0 (S 70)	0.0 (S 70)	+22.0 (S 63)	+35.6 (S 55)	−42.1 (S 63)
No control	3452.7	3731.8	4153.9	3702.1	3059.1	3942.5

[a] See Table 66

4.3.3.5 *Benefit of Infestation Forecasts*

In view of the comparatively low potential loss with an average of 509 kg per ha (Table 72) which corresponds to 11% of the potential yield, the opportunity costs of cash have not been taken into account in these computations. The alternative use of the maximum amount budgeted for crop protection (230 Pesos) would enable a yield increase of a magnitude corresponding to this crop loss to be realized.

Table 72 again gives the best strategy for each infestation event in resistant varieties. Under the economic conditions of the rainy season, the average net return to these strategies amounts to 3957 Pesos ha^{-1}, so that it is only 20 Pesos higher than the strategy which on the average is best (S 9).

Looking at the individual data, it will be noted that in 40% of all cases, control is not economically feasible. In 47% of all cases, the correct decision would be S 9 and in the remaining two infestation situations, it would be S 7.

With a maximum residual loss of 344 kg ha^{-1}, all of the strategies conform to the minimum yield restriction. The potential possibilities of improving profitability by increasing control effectiveness are correspondingly low. With a potential loss of 509 and a residual loss of 180 kg ha^{-1}, the effectiveness of control amounts to as much as 64%, i.e. to the level achieved in insecticide trials. The corresponding data for non-resistant varieties are not of great usefulness, owing to the low infestation intensity. Under the given conditions, the average net returns obtained from the various best strategies amount to 3856.7 Pesos ha^{-1} (Table 73) and are thus 34 Pesos higher than the net returns based on the strategy with the maximum average net returns (S 6) (Table 70). All the strategies conform to the set cost and minimum yield restrictions. The residual

ha^{-1}, resistant varieties, Iloilo, rainy and dry season

Infestation events[a]									
7	8	9	10	11	12	13	14	15	$\bar{x}$
−212.0	−32.0	−207.0	−240.8	+195.9	−180.4	−225.2	−167.4	−151.2	−125.0
−171.2	+162.2	−84.6	−260.5	+603.4	−85.0	−322.9	−212.6	−164.4	−3.3
−187.0	0.0	−176.1	0.0	−195.1	0.0	0.0	0.0	−458.3	−98.4
(S63)	(S70)	(S63)	(S59)	(S59)	(S70)	(S70)	(S70)	(S62)	
4132.1	3828.1	4047.4	4329.0	3273.3	4118.8	4435.1	4274.1	4176.2	3982.1
−96.9	+57.5	−96.7	−113.6	+246.2	−67.6	−93.1	−46.9	−36.8	−4.3
−9.2	+292.2	+68.7	−80.9	+679.2	−73.9	−134.3	−38.4	+1.0	−142.9
−94.3	0.0	−73.8	0.0	−127.8	0.0	0.0	0.0	−183.7	−36.1
(S63)	(S70)	(S63)	(S70)	(S59)	(S70)	(S70)	(S70)	(S62)	
3924.8	3651.3	3848.6	4101.9	3152.1	3912.8	4197.3	4052.5	3964.4	3789.9

loss of 336 kg ha^{-1} corresponds to 54% of the potential loss. The average effectiveness of control is, therefore, lower than in resistant varieties.

Studying the individual data, we find that in 50% of all cases, S 6 represents the best strategy. In one case, the best measure from an economic point of view would be S 15, in two further cases, it would be non-control.

As regards the economic feasibility of crop protection measures, we can say that reliable infestation forecasts would not result in any major incremental benefits even if the control effectiveness of the available measures were to be improved. However, this should not automatically lead us to conclude that a less costly prophylactic measure should be recommended, since, in the case of routine low-dosage applications, long-term aspects must also be taken into account. Owing to the increased danger of a build-up of pest resistance to the pesticides, the effectiveness of the latter may decrease in the long term and the costs of crop protection would rise. Under these conditions, the alternative use of the limited cash funds available would increase in value and as a result, the well-targeted application of pesticides would become more important.

4.3.4 Comparative Assessment of Model Results for the Three Investigation Areas

4.3.4.1 Economics of Crop Protection Measures

The results arrived at for the individual investigation areas show that, depending on the assumptions made, different strategies must be regarded as optimum. The infestation events and the resulting crop losses are the main determinants of

Table 72. Economic parameters of superior control strategies, resistant varieties, Iloilo[a]

Economic parameters	Infestation events[b]															
	1	2	3	4	5	6	7	8	9	10	11	12	13	14	15	$\bar{x}$
Returns ($P\ ha^{-1}$)	3838.9	3911.9	4153.9	3797.2	3764.7	3942.5	3924.8	3943.5	3917.3	4101.9	3831.3	3986.7	4197.3	4052.5	4000.2	3957.6
Yield ($kg\ ha^{-1}$)	4402.0	4397.5	4466.5	4318.6	4364.8	4239.9	4220.2	4502.2	4444.2	4410.6	4432.9	4519.6	4513.3	4357.5	4473.3	4404.2
Costs of pest control	184.6	138.4	0.0	184.6	184.6	0.0	0.0	184.6	184.6	0.0	184.6	184.6	0.0	0.0	138.4	–
Residual loss ($kg\ ha^{-1}$)[a]	182	187	118	265	219	344	364	82	140	173	152	65	71	226	111	180
Potential loss ($kg\ ha^{-1}$)[a]	871	572	118	603	1295	344	364	658	446	173	1195	377	71	226	321	509
Strategy no.[a]	S9	S7	S70	S9	S9	S70	S70	S9	S9	S70	S9	S9	S70	S70	S7	–

[a] Cf. Table 56
[b] See Table 66

Table 73. Economic parameters of superior control strategies, non-resistant varieties, Iloilo[a]

Economic parameters	Infestation events[b]						
	1	2	3	4	5	6	$\bar{x}$
Returns ($P\ ha^{-1}$)	3764.7	2992.8	4111.1	4087.3	4191.8	3992.4	3856.7
Yield ($kg\ ha^{-1}$)	4182.0	3463.7	4554.0	4529.6	4507.6	4293.9	4255.1
Costs of pest control ($P\ ha^{-1}$)	92.3	181.7	92.3	92.3	0	0	–
Residual loss[a] ($kg\ ha^{-1}$)	402.0	1163.0	30.0	55.0	77.0	291.0	336.0
Potential loss[a]	718.0	1577.0	344.0	369.0	77.0	291.0	619.0
Strategies	S6	S15	S6	S6	S70	S70	–

[a] Cf. Table 56
[b] See Table 67

the economic results to be obtained from a specific strategy. In addition, the ratio between the costs and the control effectiveness of a specific strategy is also of decisive importance. The costs of the control strategy must in turn be viewed in relation to both the value of the alternative use of cash and the maximum cash funds available for crop protection. As Table 74 shows, the optimum strategy also depends on the decision rule employed. Using the Laplace rule in conjunction with the minimum yield and liquidity restrictions discussed in detail above, higher average net returns will be arrived at for all three areas than by basing decisions on the extremely pessimistic maximin rule. Furthermore, it must be borne in mind that the latter decision rule always calls for crop protection strategies involving higher costs. A further distinction results from the opportunity costs of cash which are highest in Camarines Sur and lowest in Iloilo. If the opportunity costs are taken into account, preference tends to be given to less costly crop protection measures; in the extreme case (Iloilo) even the decision to dispense with control can be plausible. Therefore, including the opportunity costs will perforce lead to a better use of the variable production factors (Headley 1972).

In the row "marginal net returns" in Table 74, the differences between the optimum strategy and non-control net returns are listed. The opportunity costs of cash also play a decisive role: when the marginal yield resulting from the cash funds released is higher than the prevented loss of the most successful strategy, then control is no longer worthwhile. The marginal yields of crop protection are consequently always lower when production alternatives are taken into account. The highest marginal net returns, averaging over 600 Pesos ha^{-1} are obtained for the Nueva Ecija investigation area. In Iloilo, on the other hand, the average marginal net return amounts to only 26 Pesos ha^{-1} when opportunity costs are not taken into account.

The average and minimum yields computed apply only when there is an equal probability of infestation events occurring in resistant and non-resistant vari-

Table 74. Economic parameters of crop protection for the three investigation areas

Selection rule infest. event	Parameter	Unit	Nueva Ecija		Camarine Sur		Iloilo	
			With opport. costs[a]	Without opport. costs	With opport. costs[a]	Without opport. costs	With opport. costs[a]	Without opport. costs
Optimum strategy acc. to Laplace rule (resistant and non-resistant varieties)	Strategy no.		S15	S15	S6	S15	S70	S6
	Control costs	$P\ ha^{-1}$	159.3	159.3	89.1	155.4	0	92.3
	Average net returns	$P\ ha^{-1}$	4388	4319	3503	3286	3942	3798
	Marginal net returns[b]	$P\ ha^{-1}$	613	660	332	412	0	26
Optimum strategy acc. to max/min rule (resistant and non-resistant varieties)	Strategy no.		S29	S29	S6	S15	S6	S23
	Control costs	$P\ ha^{-1}$	314.0	314.0	81.6	288.6	92.3	181.7
	Average net returns	$P\ ha^{-1}$	4277	4277	3260	3260	3934	3710
	Marginal net returns[b]	$P\ ha^{-1}$	502	618	89	386	0	0
Optimum strategy acc. to Laplace rule (resistant varieties)	Strategy no.		S9	S9	S70	S9	S70	S9
	Control costs	$P\ ha^{-1}$	182.6	182.6	0.0	178.1	0	184.6
	Average net returns	$P\ ha^{-1}$	5053	4988	4456	3922	3982	3933
	Marginal net returns[b]	$P\ ha^{-1}$	201	347	0	188	0	143
	Potential loss	$kg\ ha^{-1}$	–	606	–	761	–	509
Optimum strategy acc. to Laplace rule (non-resistant varieties)	Strategy no.		S29	S29	S36	S36	S6	S6
	Control costs	$P\ ha^{-1}$	314.0	314.0	288.6	288.6	92.3	92.3
	Average net returns	$P\ ha^{-1}$	4096	4096	2981	2981	3947	3774
	Marginal net returns[b]	$P\ ha^{-1}$	1398	1419	674	968	21	34
	Potential loss	$kg\ ha^{-1}$	–	2110	–	2447	–	619

[a] The opportunity costs of cash correspond to a yield increase of 9.72% (Nueva Ecija), 14.52% (Camarines Sur) and 11.4% (Iloilo) respectively per unit of money (Peso)

[b] Compared to non-control

eties. On the other hand, considering the results for resistant and non-resistant varieties separately, we find – as expected – that the economic feasibility of chemical control measures is definitely lower in resistant varieties. If, in the former case, opportunity costs are taken into account, then – taking into account all the infestation events a control will on the average cease to be economically feasible except in Nueva Ecija. Excluding the opportunity costs of cash, control will result in an incremental return ranging from 143 to 347 Pesos ha^{-1}, corresponding to a cost/benefit ratio of 1:2.9 (Nueva Ecija), 1:2.1 (Camarines Sur) and 1:1.8 (Iloilo). Comparing these ratios with the generally accepted threshold ratio of 1:2 for the acceptance of a new technology (Ruthenberg 1977), we will find that it is only in the Nueva Ecija area that the threshold value is definitely exceeded, although the high average crop price in the 1980 rainy season changed the above results in favour of chemical crop protection. With the exception of the Iloilo area, the economic feasibility of control is clearly higher in non-resistant varieties. Here, the opportunity costs of cash play a considerably smaller role because of the high average potential losses amounting to over 2 t ha^{-1}. In Nueva Ecija, the marginal net returns to a control measure are higher than in Camarines Sur and the potential losses are lower. This is a result of the different effectiveness of control.

The economic feasibility of crop protection measures therefore depends to a great extent on the probability of varietal resistance breaking down. The spread of multiple resistant varieties, which in the Philippines are already used on 80% of the rice acreage (Kush and Chandary 1981) may help to reduce the probability of new pathogen races and pest biotypes developing before new varieties are bred, provided the preconditions for long-term resistance are created at the same time. In this case, with the input cost/crop price ratio remaining unchanged, the economic feasibility of chemical crop protection in irrigated rice cultivation would become lower in future. This statement does not apply to weed control since here, different factors are involved.

4.3.4.2 Benefit of Information Improvement

In the case of warnings issued by the official crop protection service, we must distinguish between negative warnings, i. e. information signalling that varietal resistance is likely to continue, and positive warnings, i.e. information signalling that varietal resistance will break down. It will be noted that the benefit – expressed in terms of the difference in average net returns between the initial strategy and the measure recommended by the warning service – depends more on the type of warning than on the investigation area concerned.

In almost all cases, a warning in the positive sense results in a greater incremental return than a negative warning (Table 75). In Nueva Ecija and Camarines Sur, the incremental return amounts to over 420 Pesos ha^{-1} and is thus higher than in Iloilo, where it only amounts to 212 Pesos ha^{-1}, although infestation intensity is comparatively low in nonresistant varieties. In these figures, opportunity costs are not included. If opportunity costs are taken into account, the results will change; when the cost of the control strategy replaced is substantially below the liquidity limit and the value of the alternative use of the

Table 75. Economic parameters of different levels of information improvement in chemical crop protection in the three investigation areas

Parameter	Type of warning[a] / inf. situation	Initial situation	Nueva Ecija		Camarines Sur		Iloilo	
			With op. c.	Without op. c.	With op. c.	Without op. c.	With op. c.	Without op. c.
Potential benefit of a warning (P ha^{-1})	Negative	Best farmer strategy of the investigation area concerned	+182	+200	+269	+162	+125	+166
	Positive		+378	+427	+161	+427	+255	+212
	Negative	Optimum strategy acc. to Laplace rule	+290	+303	+682	+173	0	0
	Positive		+82	+142	+ 22	+ 99	+ 21	+266
Necessary minimum reliability of a warning (%)	Negative	Best farmer strategy of the investigation area concerned	87	70	55	82	0	0
	Positive		50	45	45	65	0	0
	Negative	Optimum strategy acc. to Laplace rule	85	90	60	85	20	63
	Positive		35	10	20	35	80	27
Benefit of control thresholds (P ha^{-1})	—	Best farmer strategy of the investigation area concerned	+28	−36	−283	−152	−31	−12
	—	Optimum strategy acc. to Laplace rule	+120	+66	−32	+21	−91	−178
Benefit of infestation forecast (P ha^{-1})	Res. varieties	Best farmer strategy of the investigation area concerned	+247	+198	+104	+183	+245	+191
	Non-res. varieties		+481	+487	+340	+461	+357	+244
	Res. varieties	Optimum strategy acc. to Laplace rule	+355	+225	+274	+194	+120	+25
	Non-res. varieties		+185	+202	+133	+201	+123	+298
Control effectiveness of the superior strategy	Res. varieties	Average of maximum net returns achieved in the different infestation situations	39	64	32	54	22	64
	Non-res. varieties		61	70	34	57	18	46

[a] Positive = a warning signalling the breaking down of varietal resistance, negative = a warning signalling that varietal resistance will continue
[b] op. c. = opportunity costs

cash funds saved is correspondingly high per unit of money, the benefit will decrease; on the other hand, when costs remain the same or decrease, and the alternative use of cash results in an incremental return, the benefit will increase.

The benefit of positive warnings consists primarily in reducing the crop losses which would have occurred in the case of a mass reproduction of pests. Negative warnings, on the other hand, give rise to low-cost strategies since the lower potential loss calls for a less intensive pesticide input. The costs of the strategies recommended by the official extension service may all the same still exceed the costs of the best farmers' measures. This can again be ascertained by comparing the alternatives with and without opportunity costs. Whenever net return drops after opportunity costs are taken into account, it means that the costs of the recommended (best) strategy exceed the costs of the initial strategy (best farmers' strategy concerned).

Including opportunity costs, the benefit of a negative warning is highest in Camarines Sur, amounting to almost 270 Pesos ha^{-1} (Table 75). Without opportunity costs, it is highest in Nueva Ecija, where it amounts to 200 Pesos. Here, a less costly measure is replaced by a more expensive one, since in resistant varieties the latter will lead to a higher cost-output difference than the initial strategy which was geared to both situations.

In the case of negative warnings, the benefit is compared more with the optimum strategy governing both types of events than in comparison with the best farmers' strategy. In the case of positive warnings, the resulting benefit is lower since the profitability of the computed optimum strategy is already high (Table 74).

Owing to the different consequences resulting from incorrect information, the positive type of warning must possess a considerably higher degree of reliability, amounting to at least 60–90%. Furthermore, due to the great difference between the optimum strategies for E_1 and E_2 in Nueva Ecija, the degree of reliability of the warning must be higher in the latter area than in Camarines Sur. The assumption that the farmer actually bases his decisions on the warnings, raises the question whether the computed minimum reliability corresponds to the individual farmers' assessment of the risk involved. Assuming the warning has a reliability of 90% – provided there is actually a possibility of varietal resistance breaking down – the opposite event would occur once in ten cropping periods. If the situation is uncertain, the probability of E_1 or E_2 occurring corresponds to exactly 0.5. On the assumption of a 10% probability of the warning being incorrect, the undesired event would thus occur once in 20 cropping periods. With a error probability of 0.1, it would already occur once every 5 years. Although it is not possible to determine a generally valid limit, it is probable that the latter situation would no longer be acceptable from the point of view of the farmer. However, if progress is made in breeding resistant varieties and the probability of varietal resistance breaking down can be reduced, then a lower degree of reliability would suffice for the warnings, and at the same time, the negative type of warning would also be justified.

Compared with the benefit resulting from the best farmers' strategy, the benefit to be derived from the current control thresholds is marginal or even negative. This is due to the fact that the effectiveness of the threshold concept is lower when the threshold level is reached, and that in all the areas there is a comparatively large number of cases in which the threshold is reached by one or several pests. If opportunity costs are taken into account, then the farmers' strategies are superior to the control threshold concept in all three areas. This is due to the fact that when two pests are involved, the costs of control on the basis of thresholds are higher than the costs of the farmers' strategies. The development of multiple thresholds and the contemporaneous improvement of control effectiveness are consequently important preconditions for the economic superiority of this concept.

The final level of information improvement, i.e. achievement of an invariably correct infestation prognosis, still comprises considerable reserves: in the case of resistant varieties, the possible incremental returns range from 100 to almost 250 Pesos ha^{-1} (Table 75); in the case of non-resistant varieties, they even range from over 240 to almost 500 Pesos ha^{-1}. The potential benefit is highest in Nueva Ecija since the heterogeneous infestation structure, especially in non-resistant varieties, calls for a frequent change in optimum strategy.

Furthermore, the potential benefit of infestation shows that an improvement of the current control thresholds can result in considerable incremental returns and that more intensive research work in this area would be justified.

Studying the average control effectiveness of each one of the best crop protection measures, we will find a further possibility of improving the profitability of crop protection. The corresponding kill coefficients listed in Table 75 range from 46 (Iloilo) to 70% (Nueva Ecija), and must therefore be regarded as low by the standards of more developed agricultures.

Overall, it seems that the economic feasibility of chemical crop protection in irrigated rice cropping can definitely be stepped up, both by technical improvements and by creation of a better information basis. However, the following factors impose limits on these possibilities of improvement:

- The magnitude of the potential crop loss in resistant varieties which cannot generally be regarded as high (cf. Table 74).
- A possible decline in the probability of varietal resistance breaking down before the introduction of a new variety, which would mean that the excessive use of pesticides could be dispensed with.
- The specific situation of the farmers with respect to acceptance of chemical crop protection measures, the main factors at play being the farmers' cash surplus and net incomes as well as their level of know-how.

4.3.4.3 Realizability of Strategies from the Point of View of the Individual Farm

Figure 16 is a graphic presentation showing the results of the most important strategies with respect to the set restrictions regarding minimum yields and cash input. It is interesting to note that there are no strategies either in Nueva Ecija during the dry season nor in Camarines Sur during the rainy season, both of which are above the minimum yield line and also to the left of the cash constraint line. Therefore, in these cases, no strategy conforms to the set restrictions and

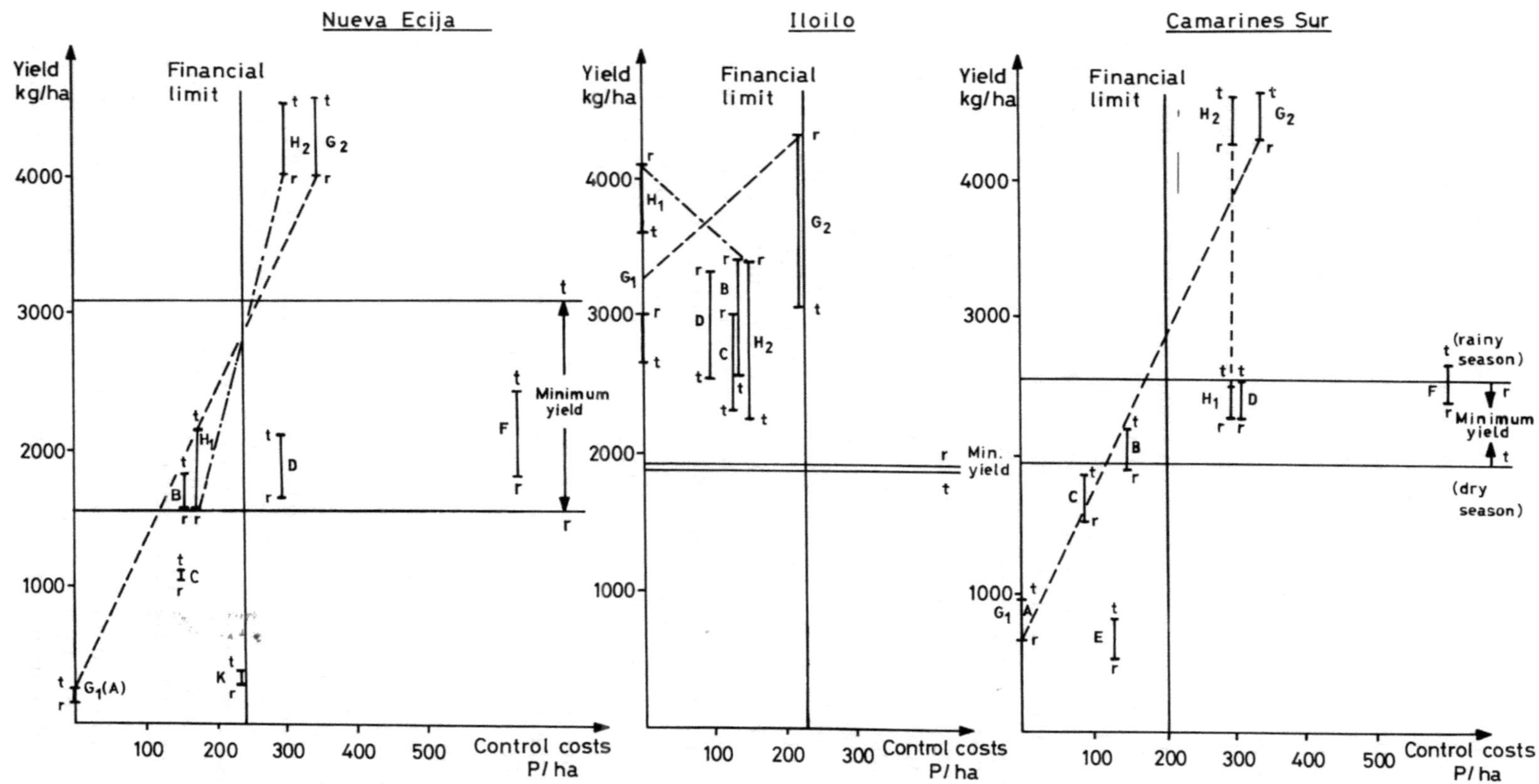

D = S 29 (2 applications of Azinophosethyl)
E = S 29 (1 early application of Diazinon)
F = S 45 (4 applications of Azinophosethyl)
K = best farmer strategy without effective tungro control

D = best farmer-strategy overall (S6)

D = S 29 (2 applications of Azinophosethyl)
E = S 7 best farmer-strategy without effectiv tungro control
F S45 (4 applications of Azinophosethyl)

A = non-control, **B** = Carbofuran, **C** = best farmer-strategy of area concerned, **G** = control threshold concept,
H = infestation forecast, **t** = dry season, **r** = rainy season ;

Fig. 16. Yields and costs of different crop protection strategies as regard to minimum requirements

thus increasingly high costs will be involved to approach the required minimum yield. Even selection of the correct strategy in each situation (infestation prognosis) would lead to minimum yields not being achieved and the liquidity restrictions being exceeded in at least one case each (Nueva Ecija). In Camarines Sur, the liquidity restrictions were exceeded in all the cases studied. As long as reliable information is not provided, regarding the breaking down of varietal resistance, the control threshold strategy will have to be regarded as equal to non-control as far as yield risk is concerned. As regards control costs, these exceed the liquidity restriction as soon as two pest species both reach their respective thresholds.

On the basis of the assumed infestation situation in non-resistant varieties, the Province of Iloilo is an exception. All the strategies, including non-control, are acceptable as far as conforming to the set restrictions is concerned. However, the respective best farmers' strategies (strategy C) violate the minimum yield restriction both in Nueva Ecija and in Camarines Sur. On the other hand, even the lowest yield clearly exceeds the required farm household consumption of rice (cf. Table 45).

The strategy with the most favourable results (except for Iloilo), both from a cost point of view as well as with respect to the risk of crop loss (minimum yield), consists of the application of Carbofuran before transplanting (strategy B). The costs of this strategy, amounting to 160 Pesos per ha are only marginally higher than the costs of the best farmers' strategy. In Nueva Ecija, the minimum yield is more than 500 kg ha^{-1} higher than the corresponding yields of the farmers' strategy, and in Camarines Sur, it is about 300 kg ha^{-1} higher (Fig. 16). This raises the question as to why this particular measure is used so seldom by the farmers. This pesticide is used for only 15 to 20% of all insecticide applications, in Camarines Sur and Iloilo for even less than 10%. It must furthermore be noted that the farmers generally do not apply this granule until after transplanting which considerably reduces its effectiveness. It is quite possible that the reason underlying prevention of this strategy is the toxicity of the pesticide which is among the most highly toxic used in rice (Magallona 1980). Applied before transplanting, it would endanger the workers during the transplanting operation.

Increased propagation or even subsidizing of Carbofuran would therefore be questionable for humanitarian reasons alone. However, this question must also be examined from an economic point of view, both on a short-term and a long-term basis. In the short run, it is necessary to investigate whether the additional expenditure on pesticides will result in adequate incremental returns. In this context, further trials are considered to be necessary (Reissig et al. 1981). As regards decreasing the risk of crop failure, the question regarding the probability of extremely high crop losses again becomes relevant.

Viewed on a long-term basis, the permanent application of Carbofuran must be seen in the light of the resulting increase in pest resistance (Venugopal and Litsinger 1980).

Taking into account both long-term and short-term aspects, it would appear that the need-based use of pesticides is most advantageous. As shown by the results of our model computations, this strategy leads to the most favourable economic results even in situations in non-resistant varieties (especially tungro infection), whilst its costs are acceptable. Naturally, timely and reliable warnings, based if possible on the data of the preceding season, are a precondition for such targeted pesticide inputs. It would then be possible to decide before each cropping season whether or not a loan or subsidy program is necessary for specific areas.

When the control threshold concept is used, a clear difference can also be noted with respect to the pesticides most commonly used by the farmers. In the case of control thresholds, the less costly pesticides such as BPMC and Carbaryl, dominate, with the recommended dosages being assumed in our computations. The farmers, on the other hand, use primarily the following groups of chemicals: Monocrotophos, Azinophosethyl and Brodan (Chlorophyrifos and BPMC) and, in one case (Camarines Sur) Endosulfan at a considerably reduced rate of application. It would appear necessary to conduct further trials to test the effectiveness of lower dosages and of the lowest economically feasible dosage of the chemicals employed by the farmers.

4.3.5 Possibilities and Practicability of the Control Threshold Concept

It is increasingly being realized in the developing countries that the prophylactic application of crop protection chemicals is not per se an answer to the existing crop disease and pest problems (Brader 1979). The systematic, calendar-based application of these pesticides has led to the destruction of the biological balance of the environment and, as a result, to the occurrence of secondary pests. In the model computations in the preceding chapters, this dynamic aspect of crop protection has been neglected.

For determining the most economical use of pesticides on the basis of control thresholds, an in-depth analysis going beyond our model computations will consequently be necessary. First of all, in view of the high level of know-how required, we must ascertain the extent to which the control threshold concept as such has any chances of being accepted by farmers in developing countries. A crop protection technology which has a lower benefit potential, but is easily understood and applied by the farmer may result in a higher overall benefit than a comparatively complicated though economically attractive technology (Herdt 1978). When considering crop protection measures, it must furthermore be taken into account that decisions regarding their implementation always have to be made on the basis of incomplete information. There is therefore a strong tendency, especially when income reserves are low, for crop protection measures to be implemented according to the safety-first principle. For this reason it is necessary, when examining the practicability of the control threshold concept, to determine the technical levels of effectiveness (control effectiveness) necessary to ensure that this concept is economically superior to the control practices commonly used to date, and to investigate at the same time to what extent pest control, based on control thresholds, is compatible with the current minimum yield and liquidity restrictions. Furthermore, the technical preconditions will have to be studied for the targeted use of pesticides on the basis of infestation monitoring. It therefore seems necessary to deal in more detail with the farmers' decision bases for the application of chemicals.

4.3.5.1 Farmers' Methods of Pest Observations and Decision Criteria

Prophylactic crop protection is often considered to be identical with routine calendar spraying, although in fact there is a substantial difference.

The results of farmer surveys regarding the decision basis for pesticide applications show that less than 10% of the farmers use routine calendar spraying as a basis for their crop protection decisions. The decision basis most frequently mentioned, on the other hand (about 50–60% of the farmers interviewed), is the observation of the pests.

In certain areas (Iloilo), damage symptoms are observed more frequently than pest populations. This may be due to the types of pests encountered most frequently. According to other surveys in rain-fed cultivation areas, two thirds of the persons questioned based their decisions on pest populations or damage symptoms (Litsinger et al. 1978).

The farmers employ various methods of observation. Frequently crops are checked in conjunction with field work (e.g. cleaning irrigation ditches). If anything out of the ordinary is noticed during such operations, the field will be inspected more thoroughly. Three quarters of the farmers interviewed are in a position to give details of the criteria used for their inspection. Intensity of infestation per hill is the most frequent criterium used (30% of the farmers interviewed). A further 19.5% base their decision on the level of infestation per unit area (m^2) and 13% examine the individual tillers (Table 76). It is surprising to note that only 1.7% of the farmers questioned mentioned the leaves as a unit of observation. On the other hand, the control thresholds for all leaf pests are expressed in terms of the relative number of infested leaves. The farmers, however, do not employ a uniform method of assessment for these pests. But it was found that the infestation thresholds on which the farmers base their decisions are generally somewhat lower (≤ 5). For the stem borer, the criteria used are largely the same, although the farmers tend to go by the overall appearance of the hill rather than by that of the individual tillers. In the case of the brown planthopper, the number of infested hills and the number of insects observed are decisive for the farmers although many are unable to specify exactly the unit of reference used. It is nevertheless clear that for this insect, the control threshold is also lower than that recommended by the extension service. Only four of the farmers questioned mentioned the green leafhopper as a vector of the tungro virus (Table 76).

As regards the observation behaviour of the farmers with respect to the implementation of chemical measures, it can be said that although the measures can indeed be regarded as prophylactic due to the low infestation thresholds used, they are definitely not calendar sprayings. From this point of view, the necessary preconditions for the use of control thresholds can therefore be said to exist. The only problem involved is the method of infestation measurement, since the farmers do not actually carry out quantitative measurements in a stric-

Table 76. Criteria used for pest observations and control thresholds by farmers and the extension service in absolute values and in %

Pest	Farmers' observation criteria																Σ	Control thresholds used by Crop Protection Service	
	Adults		Leaves		Tillers			Hills		Area (m^2)			Not clearly defined						
	≤5	>5	≤5	>5	≤5	>5 –10	≥10	≤5	>5	≤5	5–10	>10	≤5	5–10	10–20	≥20		Criterium	Level[a]
Brown plant-hopper	4	2						7	3				2	1	7	2	28	No. per hill	5 (20)
Green leaf-hopper					2											2	4	No. per 10 net sweeps	5 (2)
Stem borer					9	3	3	14	5	3	3	4	4	1	2	1	5	Percent dead hearts	15 (5)
Army worm		2			1			6		3	1	2	3	2		1	21	Percent infested leaves	25 (15)
Rice case-worm	2		1	1					1	1		2	1	1			10	Percent infested leaves	25 (15)
Rice leaf folder												1	1	1			3	Percent infested leaves	15
Rice whorl maggot										2							2	Percent infested leaves	75 (25)
Total	10		2		18			36		22			32				12	—	—
(Percent)	(8.3)		(1.7)		(15.0)			(30.0)		(18.4)			(26.6)				(100)		

[a] Figures in brackets refer to generative growth phase
Source: Author's surveys, 1980

ter sense. Determination of the absolute level of infestation intensity poses a further problem. A field in which one quarter of the leaves are infested can give the impression of being neglected, so that for "optical" reasons alone the farmers concerned will not accept it.

4.3.5.2 Farmers' Ability to Interpret Infestation Situations

Although the rice farmers are very well able to distinguish the different types of pest, it is much more difficult to grade the pests according to their damage potential. Within the scope of the questionnaire survey, the farmers were asked to name what they considered to be the most dangerous pest. As shown by Table 77, stem borer is regarded as the most important pest in the rainy season in all three investigation areas. In the dry season, the brown planthopper is given as the main damage factor in Nueva Ecija (this also includes transmission of the tungro virus) and in Iloilo both stem borer and brown planthopper. Comparing this information with the infestation data obtained from field trials, differences will be noted. It is striking in particular that neither rice leaf folder (Nueva Ecija) nor rice bug (Iloilo) are mentioned by the farmers. In non-resistant varieties, the brown planthopper is mentioned even before tungro in Nueva Ecija. This may be due to the fact that outbreaks of the brown planthopper took place recently, while tungro epidemics occurred some time ago. The frequent mention of brown planthopper in Iloilo indicates that the results of the field trials were not sufficiently representative. In the above-mentioned survey in Iloilo, on the other hand, tungro was mentioned as the most important disease, although no tungro epidemic ever occurred in Iloilo (Litsinger et al. 1978). In this context it must not be overlooked that the farmers' assessments of the major damage factors are influenced by the crop protection service and the pesticide dealers. These agen-

Table 77. Major pest named by farmers as compared to results obtained in field trials, in %

Pest	Nueva Ecija		Camarines Sur		Iloilo	
	RS	DS	RS	DS	RS	DS
Green leafhopper	9.8	6.2	20.0	34.0	–	–
Brown planthopper	–	35.4	4.0	4.3	34.0	34.7
Stem borer	54.9	12.5	52.0	23.4	41.5	36.7
Army worm	7.84	29.2	–	14.9	13.1	20.4
Rice caseworm	2.0	–	–	4.3	5.7	–
Rice whorl maggot	–	–	8.0	4.3	1.9	–
Rice bug	–	–	–	6.4	–	–
Rats	23.5	16.7	16.0	8.5	3.8	8.2
Others	2.0	0	0	0	0	0
Results of field trials						
Resistant varieties	Rice leaf folder		Stem borer		Rice bug	
Non-resistant varieties	Tungro		Tungro		—	

Source: Author's surveys, 1980

Table 78. Crop losses as estimated by farmers in %

Losses	Nueva Ecija		Camarines Sur		Iloilo	
	RS	DS	RS	DS	RS	DS
≤10	15.4	12.5	10.3	17.6	13.2	36.5
>10–25	23.1	16.7	10.3	11.8	30.2	32.7
>25–50	50.0	62.5	37.5	35.3	32.1	15.4
>50	11.5	8.3	41.4	35.3	24.5	15.4
$\bar{x}$	34.3	35.9	46.8	41.8	39.5	25.9
Results of functions (cf. Figs. 7, 8)	13		17 (8.5)[a]		15 (8.85)[a]	

[a] Computed on the basis of the lower limit of the damage coefficient
RS = rainy season, DS = dry season
Source: Computed from the author's surveys

cies tend to generalize pest outbreaks occurring in other parts of the country, with the result that the farmers regard pests they have never even observed as being important.

However, quantification of the crop losses that would occur if pest control were to be dispensed with is even more difficult than the qualitative assessment of the major pests.

The results of estimates attempted by the farmers are listed in Table 78. It must be mentioned that farmers were reluctant to reveal this information since they felt overtaxed. The author nevertheless gained the impression that they tend to include extreme situations in their estimates, i.e. they do not only base their opinion on the current situation in resistant varieties, but assume that basically the same losses could occur as in the case of the brown planthopper or the tungro virus.

The loss estimates show that the majority of the farmers in Nueva Ecija and Iloilo (rainy season) put crop losses at 25–40%, whilst in Camarines Sur, a figure of over 50% was given by the majority of the persons questioned. Differences between rainy and dry season were only mentioned in Iloilo, where losses are estimated to be almost 15% lower during the dry season. However, the farmers' estimates are still clearly higher than the figures obtained from the trials (Table 78). Looking at the average losses, it will be noted that the figures given by the farmers in Camarines Sur are highest, amounting to over 40% in both rainy and dry season. The Iloilo estimates of average losses in the rainy season are surprising, since they amount to almost 40% and are thus higher than those from Nueva Ecija, although the almost identical distribution in the top three intervals shows that opinions in Iloilo differ more than in Nueva Ecija.

Although direct conclusions regarding the farmers' crop protection strategy should not be drawn from the above figures, they do seem to indicate that the farmers attach considerable importance to crop protection as a means of safeguarding their crops. Furthermore, the figures show that the farmers tend to overestimate both the crop losses as well as the technical effectiveness of crop protection measures.

4.3.5.3 Economic Benefit of Currently Existing Control Thresholds

The economic benefit of control thresholds should be reflected in the elimination of unnecessary pesticide applications. If the control threshold concept is to be superior to all other strategies, it means that when the threshold is not reached, no measure must exist which results in a higher net return. Vice versa, when the threshold is reached, the threshold concept must result in the maximum incremental returns. It has already been explained in Section 4.2.2.4 that the control thresholds currently existing do not meet these requirements.

From the point of view of the farmers, the main requirement, apart from other criteria, is that the threshold strategy should lead to more favourable economic results than their own measures. The threshold values arrived at on the basis of theoretical considerations must prove effective under field conditions. This can only be tested by means of comparative on-farm trials.

The results of a corresponding trial series carried out in 1980 and 1981 on a total of 58 farms are presented in Fig. 17. The graph shows the cumulated frequency distribution of the marginal net returns resulting from the control threshold concept and farmers' measures, as compared with non-control for two investigation areas and a total of three cultivation periods. In the case of the control threshold concept, as shown by Fig. 17, only seven results are situated in the positive or negative range of net returns respectively, while all the remaining values are located exactly on the zero line. This means that the control threshold was reached in only seven cases. With respect to the economic results to be attained by means of the control threshold concept, it is therefore clear that the probability of either incremental returns or additional losses as compared to non-control is comparatively low, amounting to only 0.1. Highest marginal loss and highest marginal net returns are almost equal, amounting to 375 and 420 Pesos ha^{-1} respectively, but on five farms the control threshold concept resulted in an incremental return whilst an additional loss occurred only on two farms. This substantiates the results of a 5-year trial series in the Province of Pangasinan, according to which the control threshold concept did not always result in returns higher than those based on non-control (IRRI 1981).

On the other hand, under the infestation conditions prevailing at the time the trials were carried out, the control threshold concept yielded better results than the farmers' measures. In the case of the latter, only a 0.4 probability of an incremental return is realized. The marginal yields and losses arrived at, amounting to almost 600 and 2000 Pesos ha^{-1} respectively, also include extreme values which can hardly be attributed to the differences in the infestation situation or in the effect of a control measure.

Considering only the average results of the three strategies – namely non-control, control thresholds and farmers' measures – on all the farms, we will note that control based on thresholds leads to a marginal net return of 145 Pesos ha^{-1} (Zeddies and Waibel 1982). Considering only the cases in which the thresh-

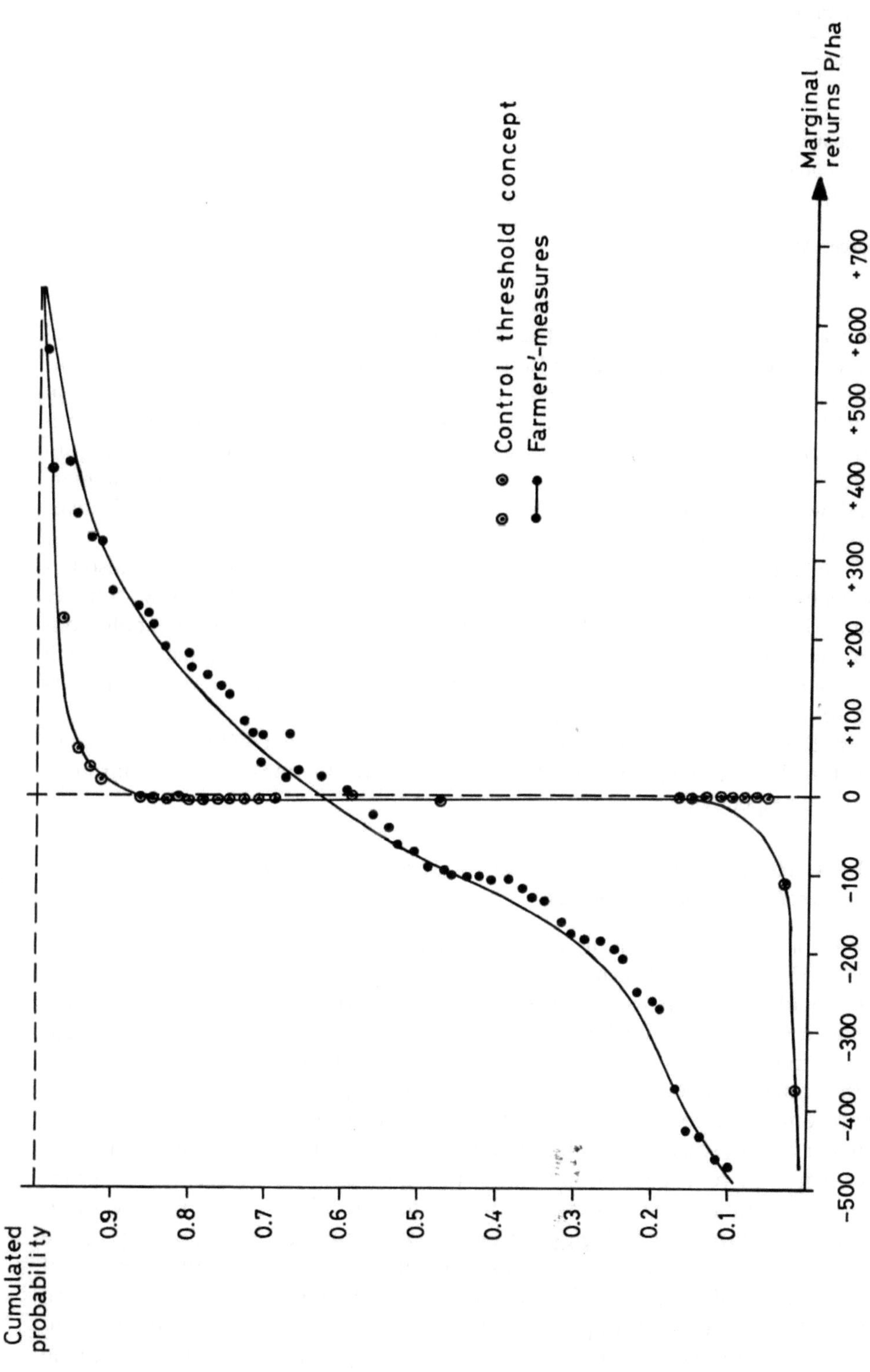

Fig. 17. Cumulated frequency distribution of marginal net returns of the control threshold concept and of farmers' strategies, in Pesos ha^{-1} Camarines Sur and Iloilo, 1980–1981

Table 79. Results of empirical tests of control thresholds

Measure is:	Control threshold was		No. of farms
	reached	not reached	
Economical	5	30	35
Not economical	2	21	23
No. of farms	7	51	58

Source: Author's computations

old was reached, we find that the difference in returns between the farmers' strategy and the control threshold amounts to 1 Peso ha^{-1}. Compared to non-control, farmers' strategy and control threshold result in a marginal net return of 43 and 42 Pesos ha^{-1} respectively. Here, the major part of the marginal net return resulting from the control threshold strategy is attributable to the savings in control costs. On the other hand, when the threshold value is reached, the marginal net return is comparatively low, so that we can conclude that the effectiveness of control is poor.

The fact that the control threshold concept is not yet fully developed also becomes evident by looking at the matter from a different point of view. Namely comparing the individual values arrived at by means of control threshold concept and farmers' strategy when the control threshold was not reached, we will find that the farmers' measure was economically feasible. i.e. better than non-control, in 21 of these cases (Table 79).

On the other hand, in two out of seven cases – as shown by Fig. 17 – the control threshold concept was economically inferior to non-control of farmers' measure respectively.

4.3.5.4 Required Reliability of Control Thresholds and Effectiveness of Control Based on Thresholds

In the preceding chapter, it was explained that the poor results achieved by means of the control threshold concept, when the threshold is surpassed by the pest population is due to the low effectiveness of control as compared with prophylactic measures. The benefit of the control threshold concept can be enhanced in two ways, the prospects of success of which must be viewed in the light of the anticipated potential loss. As long as only a marginal crop loss is expected, the recommendation of higher control thresholds will, ceteris paribus, result in a higher benefit since the number of unnecessary chemical crop protection measures would be reduced and, in years with higher crop loss, incremental net returns, as compared with non-control, would still be achieved. On the other hand, if a comparatively high crop loss is expected and the probability of the threshold being reached is also comparatively high, then the effectiveness of control achieved by means of the control threshold concept must at least reach the level of effectiveness of prophylactic measures. With the given rate of application, the latter will be a matter of choosing the optimum time for control. Due to the different infestation criteria used for determining the control thresholds, the optimum time for control is not the same for prophylactic measures as for control on the basis of control thresholds. The example of the stem borer shows that damage symptoms during the reproductive growth phase are unreliable indicators for chemical measures; at the time of damage measurement, crop loss has in part already been sustained and a measure instituted at this time may be ineffective (Venugopal et al 1982). For this reason, it is necessary to enhance the predictive nature of control thresholds so as to arrive at a higher effectiveness of control.

First of all, by means of model computations, we must determine the minimum control effectiveness required to reach the average net returns from each one of the best farmers' measures.

Following this, the author will briefly explain a possibility for arriving at realistic control thresholds.

4.3.5.4.1 Increasing the Effectiveness of Control. On the basis of empirical tests, the effectiveness of control of the threshold values currently available can be assumed to be 0.3 in the case of single pest thresholds and 0.75 in the case of multiple thresholds (provided the applications take place at short intervals) (cf. Sect. 4.2.2.4). In the simulation model, the effectiveness of control of the threshold was therefore gradually increased to find the level at which the economic results of the control threshold strategy become equal to those of the strategies which to date have proved superior. The results of these computations are shown in Fig. 18. It will be noted that there are differences between the three regions dealt with in this study.

In Nueva Ecija, preconditions for an economically feasible use of the control threshold concept are most favourable. Even a marginal increase in control effectiveness to 0.5 (single pest threshold) and 0.85 (multiple threshold), would already put the control threshold concept on a level with the best farmers' strategy. If single pest thresholds alone are improved, the average net returns from the best farmers' strategy will be reached only at a coefficient of effectiveness of 0.7; improving solely the multiple thresholds, a coefficient of 0.85 will be required before the break-even point is reached.

In Camarines Sur and Iloilo, it would be impossible to achieve net returns for the control threshold strategy equal to the level of the best farmers' strategy, even if control effectiveness were to be considerably improved, owing to the high overall level of effectiveness and clearly lower costs of the latter strategy. However, since the most favourable result of S 6 – two applications of Azinophosethyl at a greatly reduced dosage – is based on still uncertain empirical data, the next best farmers' strategy is also given in Fig. 18. If we use this strategy as a basis of comparison, the effectiveness of the control threshold strategy would have to reach a level of at least 0.6 to 0.7 in Camarines Sur, for equal net returns to be achieved, and only the simultaneous improvement of both single and multiple thresholds would lead to economic advantages.

In Iloilo, non-control would be the next best strategy. With a coefficient of effectiveness of the control threshold strategy amounting to slightly more than 0.5, the net return would be equal to non-control, whilst at 0.4, the average net re-

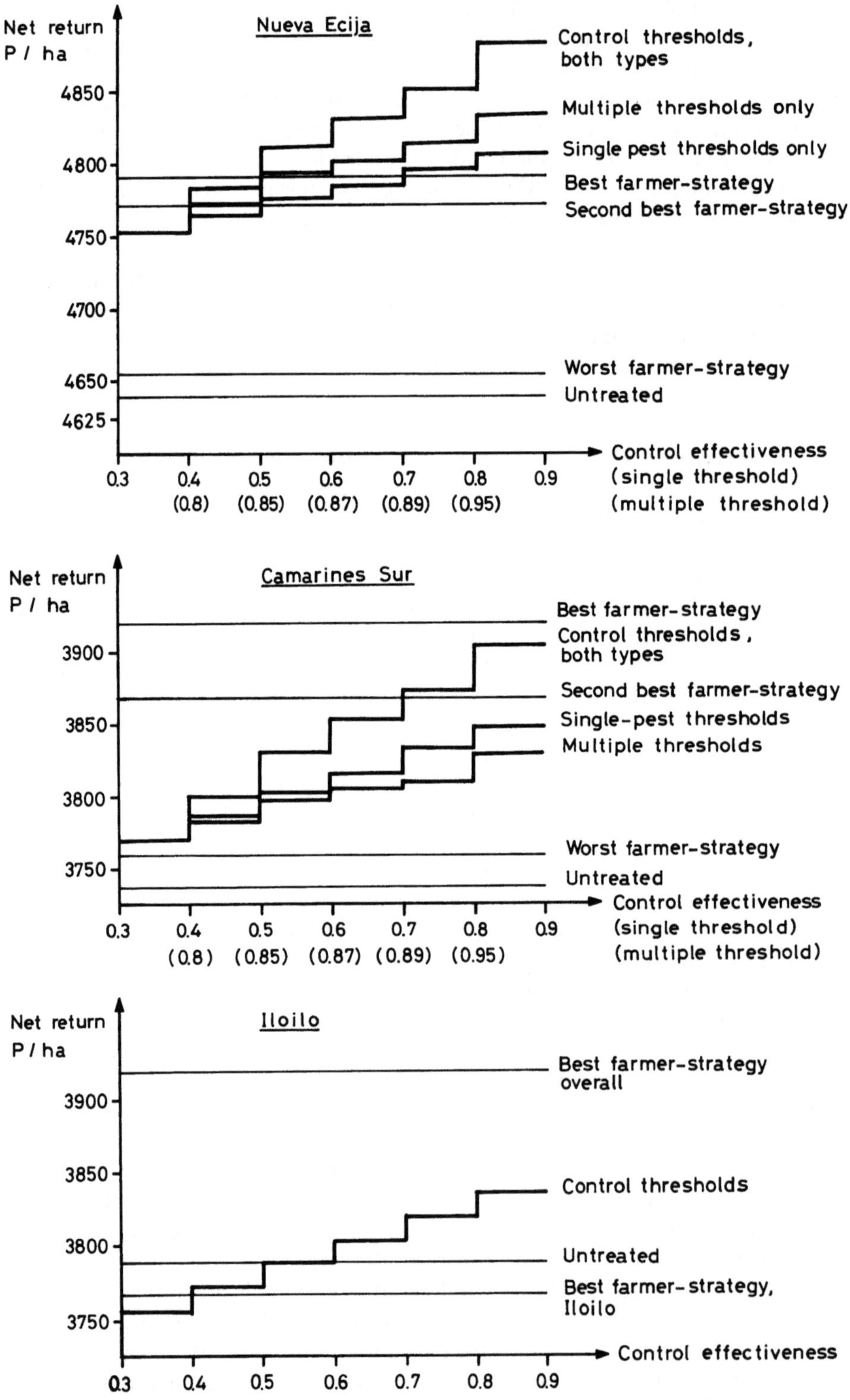

Fig. 18. Average net returns from the control threshold strategy in Pesos ha^{-1} at different levels of control effectiveness

turns would correspond to those from the best farmers' strategy in the Iloilo area.

Overall, it becomes clear that the level of effectiveness of the control threshold concept, (i.e. the relative reduction in infestation intensity when the threshold values are reached) would have to be about 50% higher than at present in order to attain the average net returns from each one of the best farmers' strategies. However, if we assess the effectiveness of the farmers' strategies on the basis of somewhat less optimistic assumptions, then even a coefficient of effectiveness of the order of 0.6 would be likely to result in economic benefits. These benefits could still be increased by putting the cash funds, thus, released to a favourable alternative use, provided the costs of the control threshold strategy correspond roughly to the costs of the farmers' measures and that there is only a slight probability of the threshold being reached.

4.3.5.5 Possibilities of Improving the Threshold Values

When the different methods of determining infestation damage coefficients were discussed (of Sect. 2.2.5), it was found that the control thresholds currently available for rice cultivation are not based on the optimum time of control. Furthermore, the infestation thresholds used are only rough estimates and do not necessarily correspond to the actual definition of the term, i.e. to break-even points for the use of pesticides. The correct values will first have to be calculated for the specific conditions concerned by means of the formula developed by Norton (1976) which has been referred to previously.

The values thus computed will guarantee that pesticides are used more sparingly. When the overall infestation intensity is low, this would also offer economic advantages due to the fact that when the threshold is reached, control costs would at least be covered. Thus, control thresholds computed by taking into account the respective economic conditions will represent a first improvement. In Table 80, the computed values are listed according to areas. It will be noted that the threshold values used by the official extension service for pests such as rice whorl maggot and rice leaf folder are still too high, although there are differences between the three cultivation areas. In Iloilo, as a result of low crop prices and higher interest rates, the threshold values are highest; in Nueva Ecija they are lowest due to the economic conditions prevailing in that area. For the army worm occurring together with other leaf pests, particularly stem borer and rice bug, the threshold values of the official extension service are too low. In the case of the stem borer, the values given in brackets refer to the damage coefficients which are computed from field trials and are considerably lower than those based on glasshouse trials. If the former value is used as a basis, then in view of the rather high control costs, the control threshold would have to be increased to a value of over 10% dead hearts, and in Iloilo even to over 20% (Table 81). Under field conditions, it would therefore be highly unlikely that this threshold would be reached. Although compared with the farmers' "rules of thumb", these computed values would on an average increase the economic feasibility of control, this is not necessarily always the case. The values computed are single pest thresholds which assume that only one pest species is in-

Table 80. Single pest control thresholds computed by the author as compared to the threshold levels used by the extensive service[a]

Pest		Threshold values						
		Nueva Ecija		Camarines Sur		Iloilo		Extension service[b]
		RS	DS	RS	DS	RS	DS	
Rice whorl maggot[c]	310	48	42	55	54	59	59	75
Army worm, etc. leaf feeding pests[c] (veg. phase)	110	19	21	23	24	24	24	25
Rice caseworm[c] (veg. phase)	110	19	21	23	24	24	24	25
Army worm, etc. leaf feeding pests[c] (gen. phase)	110	16	13	19	18	20	28	15
Rice leaf folder[c] (gen. phase)	109	9	7	10	9	11	15	15
Stem borer[d] (gen. phase)	199	7 (15)	6 (13)	9 (18)	8 (17)	9 (19)	12 (27)	5
Rice bug[e]	114	8	7	10	9	10	15	6

[a] Assumptions for Nueva Ecija: yield = 4218 kg ha^{-1} (RS), 5531 kg ha^{-1} (DS), price = 1.27 P kg^{-1} (RS), 1.14 P kg^{-1} (DS), interest rate = 13.8% per cropping period
Assumptions for Camarines Sur: yield = 4460 kg ha^{-1} (RS), 4815 kg ha^{-1} (DS), price = 1.0 P kg^{-1} (RS), 0.95 P kg^{-1} (DS), interest rate = 11.0% per cropping period
Assumptions for Iloilo: yield = 4584 kg ha^{-1} (RS), 3250 kg ha^{-1} (DS), price = 0.93 P kg^{-1} (RS), 0.95 P kg^{-1} (DS), interest rate = 15.0% per cropping period
[b] No difference is made between rainy and dry seasons
[c] Percent infested leaves
[d] Percent dead hearts
[e] No. of bugs (m^{-2})
Source: Author's computations

volved. It will therefore also be necessary to compute multiple thresholds for the most important pest combinations which can either be controlled at the same time or for the case in which the control measure will also have an effect on damage factors occurring subsequently.

In cases where the crop loss functions include several pests as variables, a consistent computation of the multiple thresholds is possible. On the other hand, if only separately computed loss coefficients are available, possible antagonistic or synergistic interaction between two or more pests is neglected.

In the first case, we can employ the function given in Table 19 for rice leaf folder and defoliator. If, in accordance with the definiton of control thresholds, we equate the control costs with the monetary loss prevented, the control costs are given by:

$$K_0 = a + (E \cdot b_1x_1 \cdot e_1x_1 \cdot p) + (E \cdot b_2x_2 \cdot e_2x_2 \cdot p),$$

where: K_0 = control costs
a = Y intercept of loss function
b_1x_1, b_2x_2 = regression coefficient for pest 1 and pest 2 respectively
e_1x_1, e_2x_2 = effectiveness of control of pest 1 and pest 2 respectively
E = potential yield
p = crop price.

If we solve this equation using the given assumptions for price, yield and control effectiveness, we obtain pest 1 as a function of pest 2.

Using the data from Nueva Ecija (rainy season) as an example, the method of computation will be shown below:

$$124 = -0.54 + (4218 \cdot 0.007764x_1 \cdot 0.3 \cdot 1.27) + (4218 \cdot 0.00885x_2 \cdot 0.3 \cdot 1.27)$$
$$x_1 = 9.98 - 1.14x_2$$

The results shows all possible infestation combinations of defoliators and rice leaf folder which will result in the multiple threshold being reached. The expression computed corresponds to the isocline of the break-even point between cost of control and marginal gross return. This can be clearly demonstrated by means of a graph. Fig. 19 shows the multiple thresholds for rice leaf folder and defoliators , as well as for stem borer and rice leaf folder, as a connecting line between the level of infestation by pest 1, in the case of zero infestation by pest 2, and vice versa. All the infestation combinations situated within the area between the point of intersection of the coordinates and the connecting line correspond to an infestation intensity below the multiple threshold.

In the case of a 5% infestation by the rice leaf folder, a 4.3% defoliator infestation would thus suffice in Nueva Ecija for control measures to be taken. In Camarines Sur, an army worm infestation of slightly over 6% would be required and in Iloilo, an infestation of 7% (Fig. 19). In the case of a 5% stem borer infestation, an 11% infestation by the rice leaf folder would be needed in Nueva Ecija for the threshold to be reached, a rice leaf folder infestation of 13% in Camarines Sur and of 15% in Iloilo.

The multiple thresholds computed are lower than either the computed single pest thresholds or the rule-of-thumb figures used to date. This naturally applied only to the infestation combinations for which it was possible to compute a multiple threshold. It was not possible to compute multiple thresholds for the leaf pests occurring in the early stage of crop growth, since rice whorl maggot and army worm have to be controlled with different pesticides. The thresholds for rice leaf roller and army worm are identical, so that a marginal rate of substitution of 1 can be used.

The calculation of single and multiple thresholds is based on the time of control to which the current thresholds correspond and which, as already explained earlier, does not represent the optimum time of control. In the case of prophylactic measures, as opposed to the control threshold concept, the pesticides are generally already applied during the vegetative phase. In the control threshold concept, on the other hand, it is assumed that the threshold value and

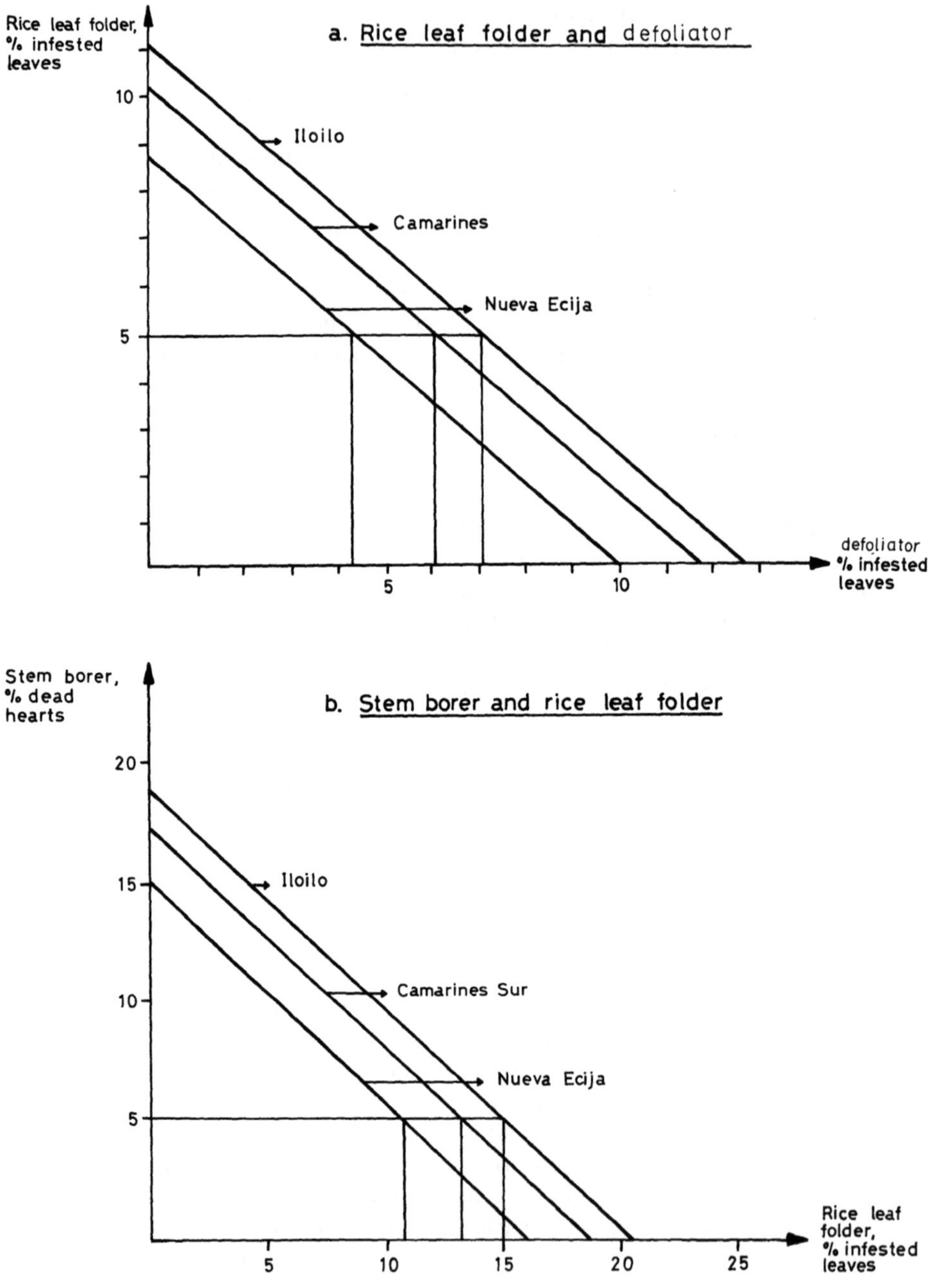

Fig. 19a, b. Multiple control thresholds for two selected pest complexes, rainy season. **a** Rice leaf folder and defoliator; **b** stem borer and rice leaf folder

Table 81. Functions for estimating infestation levels at critical growth stages

Cultivation area	Pest	Type of function	Designation of variables	*r*	*F*-value	*n*
Nueva Ecija	Stem borer	$Y = 1.97 + 0.84X^{-}$	Y = Percent white heads before the harvest X = Percent dead hearts generative phase	0.522	3.38^{-}	11
Camarines Sur	Stem borer	$Y = 0.88 + 0.784X^{***}$	Y = Percent white heads before the harvest X = Percent dead hearts generative phase	0.788	65.7**	42
		$Y = 1.439 + 0.816X^{***}$	Y = Percent dead hearts generative phase X = Percent dead hearts vegetative phase	0.809	75.78**	42
Iloilo	Army worm	$Y = 0.955 + 0.511X^{***}$	Y = Peak level of infested leaves generative phase X = Peak level of infested leaves vegetative phase	0.507	11.38*	36
	Rice leaf folder	$Y = 0.508 + 0.53X^{***}$	Y = Peak level of infested leaves vegetative phase X = Peak level of infested leaves generative phase	0.646	24.39**	36
	Stem borer	$Y = 0.502 + 0.689X^{***}$	Y = Percent dead hearts generative phase X = Percent dead hearts vegetative phase	0.657	25.82	36

$^{-}$ = Significant at $P \leq 0.1$, * = significant at $P \leq 0.05$, ** = significant at $P \leq 0.01$, *** = significant at $P \leq 0.001$

Source: Author's computations on the basis of the trial data of Chapter 2

the control measures geared to it are valid for an entire growth stage. This may be the reason for the poor level of effectiveness of measures based on the control threshold concept. Until damage coefficients relating to several shorter growth stages can be developed, a different method can be used.

At this point we will come back to the concept outlined in Chapter 3.1.3, namely the estimation of the infestation levels at a critical phase on the basis of initial or early stage infestations.

Table 81 is a listing of functional relationships – derived from data of untreated plots – between the infestation levels at the different growth stages. It is noteworthy that in some cases, significant relationships were found. In the case of the function for stem borer based on data from Camarines Sur. The statistical quality is highest, with a correlation coefficient of 0.809. The high percentage of dead hearts in the generative phase is due to the infestation peak during the vegetative growth phase, although the latter cannot be related to crop loss. By means of the functions given in Table 81, the control thresholds can be derived from the economic damage thresholds. If a control measure instituted earlier has a higher level of effectiveness, the economic damage threshold will not be identical with the value given in Table 81. For an economic damage threshold of 10% dead hearts during the generative phase, the control threshold for Camarines Sur will consequently be computed as follows:

$$10 = 1.439 + 0.816X; \qquad X = \text{dead hearts, generative phase.}$$

Solving the above equation, we obtain a control threshold of 10.5% dead hearts during the vegetative growth phase.

As regards this last step in the computation of more realistic threshold values, certain reservations must be made. In as much as the function listed in Table 81 should be regarded merely as examples for demonstration purposes, the available data is inadequate to verify the approach used. In the majority of the trials conducted to determine crop losses, systematic assessments at short, regular intervals were not carried out. It was consequently not possible to determine the correlation between the infestation levels at different growth stages for all the pests. To obtain damage coefficients of a greater degree of differentiation, it would either be necessary to construct detailed models explaining the population dynamics of the individual pests and their interactions, or else a rather complicated trial programme taking these dynamic aspects into account would have to be conducted.

Alternatively, retaining the trial concept used to date, it would suffice to increase the frequency of infestation measurement to enable the corresponding functions to be constructed. The reliability of control thresholds computed by means of this two-stage method of assessment would then have to be checked in further on-farm trials. However, the above mentioned method of determining control thresholds is only justified as long as the scientific institutions are not in a position to elaborate damage coefficients relating to several plant growth stages.

5 Conclusions

On the basis of the present study, some recommendations will be presented for improving the efficiency of crop protection measures.

In this context, it is necessary to distinguish between two main areas of improvement: first, technical progress in control strategies and second, improvement of the information basis for crop protection decisions. The systematic coordination of both areas in view of developing economically and ecologically well balanced extension programmes will finally lead to the presentation of an improved basic crop protection concept.

5.1 Technical Progress in Control Strategies

If technical progress is to be achieved in control techniques, the current trial programme for pesticide evaluation will have to be expanded. The additional trials to be included should be based on the actual tests of new pesticides and pesticide combinations. Furthermore, the official pesticide evaluation trials should be followed by an "official dosage trial" to determine the economically optimum rates of application. In these trials, the dosages actually used by the farmers should be included as trial variants.

On the basis of the above information, coupled with an evaluation of the current control recommendations and typical farmers' strategies, it will be possible to formulate more realistic alternatives. In addition to the more prophylactic type of measures, the new single and multiple thresholds computed in Chapter 4.3.5.5 can then be tested as trial variants in plot field trials on fields rented from farmers. Although trials of this kind already form an integral part of the trial programs of some of the regional crop protection offices in the Philippines, a reformulation of the trial variants seems to be necessary. The costs of most of the variants of the control alternatives used to date amount to over 500 Pesos ha^{-1} (Peralta et al. 1983) and are thus much too high with respect to both a sufficient benefit-cost ratio and the farmers' limited financial resources. Control variants which aim at complete control at individual growth stages could easily be dispensed with and instead – aside from low-cost treatments – a trial variant "complete protection" as well as an untreated control plot, could be included in the trial concept.

As a main result, these trials should supply data on the effectiveness of the variants tested in controlling the different pests. For this purpose, it will be necessary to measure infestation levels at regular intervals so that from these measurements, it will then be possible to derive the proportional reduction in infestation intensity resulting from one or several pesticide applications.

If an appropriate number of trials are conducted, the purely qualitative assessments currently used, e.g. "effective" or "non-effective", can then be replaced by quantitative parameters such as the "mean" or the "standard deviation" of the control effectiveness.

Further technical progress in control strategies can be achieved by developing and using new pesticide application techniques. The International Rice Research Institute (IRRI) is currently engaged in testing particular "root zone application" methods, also in conjunction with planting machines. However, new and improved application equipment such as row marker applicators and liquid band applicators can hardly be used by individual farmers due to the high capital costs involved in comparison with knapsack sprayers. In this context, it would seem useful to conduct further research regarding the development of new organizational forms along the lines of self-help groups.

5.2 Improvement of the Information Basis

The main information gap in crop protection – and not only in rice cultivation in the Philippines, but also in more developed countries – is the inability of farmers to correctly assess an infestation situation observed in the field. As pointed out in Chapter 4.3.5.2, the farmers have a tendency to overestimate the crop losses caused by pests; they are influenced by pest calamities experienced in the past and also have a basic difficulty in distinguishing between the different damage factors. Furthermore, they tend to overestimate the control effectiveness of chemical measures. To fill this information gap, more intensive work will be required in the field of loss assessment. If rationally based control decisions are to be reached, it will be essential to elaborate approximate damage coefficients and to make them available to the decision-makers. The knowledge of crop loss relationships coupled with information on the current infestation levels will enable realistic control decisions to be taken and thus improve the economic results of rice production. To make available information on crop loss relationships, the data elaborated in the present study can initially be used. However, as the study has shown, the damage coefficients computed still have considerable confidence limits, so that it is considered necessary for additional variables to be included.

For further research on crop loss relationships, plot field trials, in which a clear graduation of infestation levels is produced, are basically suitable. But since infestation intensities are often low, the results do not always lend themselves to evaluation by means of regression analysis. To supplement field trials, greater importance should therefore be attached to the use of the single plant method for determining damage coefficients. This method offers two advantages: first, no pesticides need to be used and second, an adequate variance in infestation intensity is generally obtained due to individual heavily infested plants, even when the average infestation intensity for a large number of plants is low. In addition, this method is more suitable than plot field trials for deriving physiological damage thresholds. Damage coefficients are only of economic value when employed for decisions regarding the use or non-use of crop protection measures. To this end, systematic and reliable infestation monitoring is necessary. Basically, infestation assessments can be performed by extension personnel or farmers. However, extension workers will only be able to carry out a limited number of field observations, and in view of the extreme heterogeneity of agronomic factors in rice cultivation, it is doubtful whether such observations would be representative. On the other hand, if the infestation assessments are carried out by the farmers, farm and field-specific factors will be more completely covered. Furthermore, a greater commitment could be expected on the part of the farmers, since they have to bear the consequences of their decisions. During many visits and talks with farmers it was noticed that they have an aversion to formal records and prefer to note down important data on their calendars. The design of a corresponding farm operations calendar would be useful. In addition to technical information, such a calendar should have sufficient space for notes. The farmers then should be urged to note down the results of their infestation assessments in specially designed calendars, so that the information is subsequently available for long-term decisions. The data collected by such "pioneer farms" should then be evaluated at regular intervals and used for updating and developing extension recommendations.

Furthermore, as part of the improvement of the information basis, the official extension service must try to bring about a change in the farmers' approach. As far as it is justified on the basis of historic infestation data, the widespread safety-first approach should be counteracted. This could be done by introducing crop insurances and emergency programs. Within the scope of the present paper it is not possible to go into details regarding the organization and effects of such programmes.

It will also be necessary to devote more room in the extension programmes to the health and environmental hazards involved in chemical crop protection. This presupposes that in passing crop protection legislation more consideration is given to cost-benefit aspects.

Finally, information improvement will also consist in simplifying the decision-finding process for the farmer and presenting it to him in a quantitative form. This can be achieved by acquainting the general extension staff and educated farmers, in training courses, with the concept of the decision matrix. In this context, strategic decision trees can be gone through step by step before the cropping season and tactical decision trees during the cropping season. These decision trees should be based on research data and on the individual assessments of experienced farmers. Thus, it may be possible to arrive at a more transparent and systematic means of extending recommendations to farmers.

In this way, the vague qualitative estimates of the participants can be transformed into quantitative assessments of the effectiveness of alternative crop protection measures and of the losses, costs and returns involved. Objective data based on field trials should always be used as a basis for comparison.

5.3 A New Basic Concept of Crop Protection

The traditional approach to crop protection is changing, in the developing countries as well. The reasons are obvious: in certain years, as this study has shown, crop protection measures based on the safety-first principle may result in economic losses. When planning crop protection programmes, more emphasis will therefore have to be made upon the collection, processing and application of data that would lead to an improvement of the decision basis. The overemphasising of prophylactic chemical measures should cease and steps should be taken to change the attitude of the warning service which too easily regards itself as an emergency unit and actually even promotes the uneconomical use of pesticides. Instead, the approach to crop protection should consist in supplementing or even replacing – if this is plausible from an economic point of view – the use of pesticides by appropriate information. To this end, the preconditions for cost-benefit considerations will have to be improved, both at macro- and micro-economic levels and account will have to be taken of economic as well as non-economic constraints, also of constraints of a cultural and psychological kind. To achieve the above, the basic concept of crop protection will have to be modified, in particular the development of extension recommendations.

To date, extension recommendations have mainly been derived from single-factor trials and then combined into a "package of technology". It was then up to the extension staff to ensure that this technology was accepted by the farmers. Although the main elements of these technologies, such as seed, fertilizer and pesticides, were adopted, this did not apply to their mode of application. The farmers tend to supplement the very generalized extension recommendation with their own observations and "trials" to render them more specific, contributing in this way to the development of a "farmer-generated" technology. This often gives rise to a feeling of frustration on the part of the extension staff, who measure their success by the extent to which their recommendations are followed. If the extension service is to be organized along modern lines, its activities will therefore have to include the collecting of data at farm level, to enable new extension programmes to be formulated and developed. The conventional crop protection strategies will thus have to be replaced by complete crop protection system (Kranz 1977). The most important output of such a system will consist of the extension recommendations derived from a decision matrix.

For preparing the decision matrix, the components of the crop protection decision system listed in Chapter 3.2.3.2 are relevant. To obtain the necessary data, further specific trials will be required in addition to the routine pesticide evaluation trials, coupled with general knowledge in the field of crop protection.

The crop protection extension recommendations must also be based on trials conducted to determine damage coefficients, physiological damage thresholds and from the latter, the control thresholds. In addition, trials for determining the economically optimum rate of application of the pesticides which have been proved to be effective in the official pesticide evaluation trials are of major importance for formulating the control alternatives (treatments). A limited number of strategies will then be selected on the basis of relevant research results and of an evaluation of the current extension recommendations and typical control strategies used by the farmers. Testing these alternatives in trials will then yield the pest control function which in its simplest form indicates the reduction in infestation intensity expressed as a percentage.

A knowledge of control effectiveness and of the damage coefficients should permit infestation situations to be assessed. By including the economic parameters and using realistic selection criteria, the optimum strategy can then be determined both for an anticipated infestation situation and for a specific infestation event. The latter will correspond to the extension recommendation for the following cropping season. In this context, the decisive assumption is the definition of the infestation events likely to occur. As long as it is not possible to compute infestation probabilities, strategies can be based on the infestation events which occurred in the past, the occurrence of which is considered possible in specific cropping situations (variety, planting date), whereby the same probabilities of occurrence will be assumed. To this end, the pest development stage at the particular time concerned will have to be assessed jointly by researchers and extension staff before extension recommendations are formulated. To obtain the above information, a system of test farms will have to be set up. This will basically be only a modification of the system of observation stations already organized by the warning service (Kaske and Waibel 1981). On these test farms, which will be regarded as demonstration and observation fields, warning service extension workers will set up simple field trials. These trials will only consist of three plots: one untreated plot, one plot on which the customary extension recommendation is tested, i.e. generally the control threshold concept, and a third plot on which the farmer should employ the treatment he considers suitable. By means of these trials, it will be possible to determine the economic feasibility of the extension recommendations as compared with the measures employed by the farmers, and the control effectiveness of these measures. At the same time, an idea of the actual infestation situation, i.e. of the situation largely unaffected by the influence of chemicals, will be obtained from the untreated control plot.

Furthermore, from the yield differences between treated and untreated plots, the residual losses occurring under field conditions can be derived. The yield data from the untreated plots together with the damage coefficients will permit the potential losses to be estimated.

From these test farms, further data will also be obtained which can be used for identifying the typical control techniques employed by the farmers and for drawing up pesticide consumption statistics. Finally, data on the use of crop inputs and on the environmental yield-determining factors which should be suitable for developing yield models, can be collected. Such models are useful for identifying causal relationships as a basis for testing or reformulating trial hypotheses within the scope of the entire crop production system (Wiese 1982).

The potential benefit of this information regarding the farmers' control strategies, their economic feasibility and the crop losses sustained will not be restricted to the development of extension programs. It would also help to create a better decision basis for pesticide registration since it would permit the economic value of individual pesticides and active ingredients to be assessed more readily.

The most important organizational change resulting from a modified basic concept of crop protection will consist in the general extension workers' participating in the warning service activities to ensure better coordination in preparing and imparting extension recommendations.

Apart from field trials, the activities of the crop protection specialists would be confined to providing technical advice for the general extension workers and organizing the necessary pest control campaigns. In order to become acquainted with the results of their trial activities, the specialist staff will participate in seminars in which the development of the newly formulated extension programmes will be demonstrated.

In this context, the concept of the decision matrix would again provide a suitable methodological basis indicating the development of a possible change in the extension recommendations and in their alternatives.

From an overall point of view, a systematic and continuous generation of information will lend a higher degree of rationality to crop protection.

Appendix

Table A1. Farmer's strategies

Strategy no.	1. Insecticide					2. Insecticide					Labor input (man-day)	Total cost of control
	Name	Dosage (kg a.i. ha^{-1})	Frequency of application	Costs (P ha^{-1})	Timing[a] (growth stage)	Name	Dosage (kg a.i. ha^{-1})	Frequency of application	Costs (P ha^{-1})	Timing[a] (growth stage)		
1	Brodan[b]	0.2	2	92.8	E						2.0	129.6
2	Brodan[b]	0.2	3	139.2	E/M						3.0	194.6
3	Azinophosethyl	0.30	2	90.3	E/M						2.0	126.9
4	Azinophosethyl	0.30	3	135.4	E/M						3.0	190.3
5	Carbofuran	0.5	1	135.0	E	Brodan[b]	ca 0.2	1		L	1.5	225.9
6	Azinophosethyl	0.2	2	60.2	E						2.0	89.0
7	Azinophosethyl	0.2	3	90.3	E						3.0	133.5
8	Azinophosethyl	0.2	3	90.3	E/M/L	MIPC	0.5	1	54.0	L	4.0	204.6
9	Azinophosethyl	0.2	4	120.4	E/M/L						4.0	178.0
10	Carbofuran	0.5	1	135.0	E	Monocrotophos	0.17	1	59.5	E/M	2.5	238.4
11	Carbofuran	0.5	1	135.0	E	Monocrotophos	0.17	1	59.5	E	1.5	238.4
12	Monocrotophos	0.17	3	178.5	E						3.0	237.2
13	Monocrotophos	0.17	2	199.0	E						2.0	158.2
14	Monocrotophos	0.17	3	178.5	E						3.0	237.2

[a] E = vegetative, M = reproduction, L = ripening stage
[b] Brodan = trade name for Chloropyrifos + BPMC

Table A2. Strategies for granules

Strategy no.	Insecticide 1					Labor input (man-day ha^{-1})	Total cost of control
	Name	Dosage (kg a.i. ha^{-1})	Fre-quen-cy	Cost (P ha^{-1})	Timing[a] growth stage		
15	Carbofuran	0.5	1	135.0	P	0.5	159.3
16	Carbofuran	1.0	1	270.0	P	0.5	312.9
17	Carbofuran	1.0	1	135.0	A	0.5	159.3
18	Carbofuran	1.0	1	270.0	A	0.5	317.9
19	Carbofuran	2.0	1	540.0	E	0.5	620.2
20	Carbofuran	0.5	4	540.0	E/L	2.0	637.3
21	Carbofuran	1.0	2	540.0	E/L	1.0	620.2
22	Carbofuran	1.0	3	810.0	E/L	1.5	938.7
23	Diazinon	1.0	1	153.0	E	0.5	179.8
24	Diazinon	1.0	1	153.0	L	0.5	179.8
25	Diazinon	1.0	2	306.0	E/L	1.0	358.9
26	Lindane	1.0	1	96.0	E	0.5	114.9
27	Lindane	1.0	1	96.0	L	0.5	114.9
28	Lindane	1.0	2	192.0	E/L	1.0	229.8

[a] P = prior to transplanting
A = after transplanting
E = vegetative
L = ripening phase

Table A3. Strategies for sprays

Strategy no.	Insecticide 1					Labor input (man-day ha^{-1})	Total cost of control
	Name	Dosage (kg a.i. ha^{-1})	Frequency	Cost (P ha^{-1})	Timing[a] growth stage		
29	Azinophosethyl	0.75	2	226.0	E	5.0	314.1
30	BPMC	0.75	2	168.0	E	5.0	248.1
31	Carbaryl	1.5	2	172.2	E	5.0	252.9
32	Brodan[b]	0.75	2	359.8	E	5.0	466.4
33	Endosulfan	0.75	2	201.6	E	5.0	286.3
34	MIPC	0.75	2	162.0	E	5.0	241.3
35	Monocrotophos	0.75	2	570.0	E	5.0	705.6
36	Monocrotophos	0.3	2	228.0	E	5.0	316.4
37	Azinophosethyl	0.75	2	226.0	E	5.0	314.4
38	BPMC	0.75	2	168.0	E	5.0	248.1
39	Carbaryl	1.5	2	172.2	E	5.0	252.9
40	Brodan	0.75	2	359.8	E	5.0	466.4
41	Endosulfan	0.75	2	201.6	E	5.0	286.3
42	MIPC	0.75	2	162.0	E	5.0	241.3
43	Monocrotophos	0.75	2	570.0	E	5.0	705.6
44	Monocrotophos	0.3	2	228.0	E	5.0	316.4
45	Azinophosethyl	0.75	4	452.0	E/L	10.0	628.2
46	BPMC	0.75	4	336.0	E/L	10.0	496.2
47	Carbaryl	1.5	4	344.4	E/L	10.0	505.7
48	Brodan	0.75	4	703.6	E/L	10.0	914.5
49	Endosulfan	0.75	4	403.2	E/L	10.0	572.6
50	MIPC	0.75	4	324.2	E/L	10.0	482.5
51	Monocrotophos	0.75	4	1140.0	E/L	10.0	1411.1
52	Monocrotophos	0.3	4	456.0	E/L	10.0	632.7

[a] E = vegetative, L = ripening phase
[b] Brodan = trade name for Chloropyrifos + BPMC

Table A4. Control threshold strategy and control based on the growth stage

	Insecticide 1					Insecticide 2					Insecticide 3							
Strategy no.	Name	Dosage (kg a.i. ha^{-1})	Frequency of applic.	Cost (P ha^{-1})	Timing[a]	Name	Dosage (kg a.i. ha^{-1})	Frequency of applic.	Cost (P ha^{-1})	Timing[a]	Name	Dosage (kg a.i. ha^{-1})	Frequency of applic.	Cost (P ha^{-1})	Timing[a]	Labor input (man-day ha^{-1})	Total cost of control	Pest[c]
53	Monocrotophos	0.75	3	855	E	Brodan[b]	0.75	2	348	M	Cararyl	1.5	2	172.2	L	17.5	1705.2	
54	Monocrotophos	0.75	3	855	E	Brodan[b]	0.75	2	348	M						12.5	1460	
55	Carbaryl	1.5	2	172.2												5.0	252.8	CW/RB
56	Brodan[c] + Carbaryl	1.5/ 0.75	2	260.1												5.0	352.9	DEFL/SB
57	Monocrotophos	0.75	1	285												2.5	352.8	WM
58	Carbaryl	1.5	1	86.1												2.5	126.4	DEFE
59	Carbaryl	1.5	1	86.1												2.5	126.4	CW
60	Carbaryl	1.5	1	86.1												2.5	126.4	DEFL
61	BPMC	0.75	1	84.0												2.5	124.0	LF
62	Brodan	0.75	1	174												2.5	226.4	SB
63	Carbaryl	0.5	1	86.1												2.5	124.0	RB
64	Brodan[b] + BPMC	0.75/ 1.75	2	268												5.0	361.9	LF/SB
65	Carbaryl + BPMC	0.75/ 1.75	2	170.1												5.0	240.5	DEFE/LF
66	Carbaryl + BPMC	1.5/ 0.75	2	260.1												5.0	352.9	DEFE/SB
67	Brodan[b] + Carbaryl	0.75/ 1.5	2	260.1												5.0	352.9	CW/SB
68	Brodan[b] + Carbaryl	0.75/ 1.5	2	260.1												5.0	352.9	SB/RB
69	Brodan[b] + Monocrotophos	0.75/ 0.75	2	459.0												5.0	579.2	WM/SB
70	No control	0.0	0	0.0												–	0.0	–

[a] E = vegetative phase, M = reproductive phase, L = ripening phase; [b] DEFE = defoliator, early infestation, DEFL = Defoliator, late infestation; [c] Brodan = trade name for Chloropyrifos + BPMC

References

Alfonso PJ, Sumangil JP (1977) Control of Rice Field Rats. In: Rice production manual, IRRI (revised edn)

Anderson J, Dillon L, Hardaker B (1977) Agricultural decision analysis. Iowa State Univ, Ames

Barlow G, Jajasuriya S, Price EC (1983) Evaluating technology for new farming systems: case studies from Philippine rice farms. IRRI

Barr BA, Koehler CS, Smith RF (1981) Crop losses rice, field losses due to insects, diseases, weeds and other pest. UC/AID Pest management and related environment, protection project. Univ California, Berkeley

Bartsch R (1976) Die Ökonomik der chemischen Schädlingsbekämpfung im bewässerten Baumwollanbau in der Gezira, Sudan. Diss Hohenheim

Bernstein RH (1981) Evaluating integrated pest management in developing countries: Indonesian perspective. Background paper prepared for the tenth session of the FAO/UNEP panel of experts on integrated pest control. Rome, March 23–27, 1981

Brader L (1979) Integrated pest control in the developing world. In: Annu Rev Entomol 24: 225–254

Bradner R, Fletcher KE (1974) Insect infestation and their effect on the growth and yield of field crops. In: Bull Entomo Res 64:141–160

Brandes W, Woermann E (1969) Landwirtschaftliche Betriebslehre. Berlin, Hamburg Parey, Vol 1

Bühlmann H, Loeffel H, Nuevergelt E (1975) Entscheidungs- und Spieltheorie, Springer, Berlin Heidelberg New York

Bullen FT (1970) Benefit cost analysis of various degrees of crop protection. In: Proc Ecol Soc Aust 5:63–75

Bureau of Agricultural Economics (1974) Crop Statistics. Minist Agric, Manila

Bureau of Agricultural Economics (1980) Prices received by farmers. Minist Agric Manila

Bureau of Agricultural Extension (1972–1979) Iloilo City, Cabanatuan City and Naga City. Jahresber Masagana-99 Program

Burleigh JR (1980) Experimental design for quantifying disease effect on crop yield. In: Crop loss assessment, proc Stakmann EC, Commemorative Symp, Minnesota

Carlson GA (1970) A decision theoretic approach to crop disease prediction and control. In: Am J Agric Econom 52:2

Cendena SM, Calora FB (1964) Insect Pest of Rice in the Philippines. In: The major insect pest of the rice plant. IRRI

Central Bank of the Philippines (1981) Stat Bull, Manila

Chandler RF (1979) Rice in the tropics. A guide to development of national programmes, Westview press Boulder, Colorado

Chelliah S (1979) Insecticide application and brown planthopper resurgence in rice. IRRI

Chiang HC (1979) A general model of the economic threshold level of pest populations. In: FAO Plant Prot Bull no 27:3

Chiang HC (1982) Factors to be considered in refining a general model of the economic threshold. Entomophaga 27, Special issue

Christ DH, Lever RJAW (1969) Pests of rice. Longmans London

Cramer H (1967) Pflanzenschutz und Welternte. Pflanzenschutz-Nachr Bayer 20:1–523

De Datta SK (1977) Weed control in rice in south east east asia: Methods and trends. In: Philipp Weed Sci Bull 4:39–65

De Datta SK, Carcia FV, Chatterjee AK, Abilay WP, Alcantara JM, Cia BS, Jerera HC (1979) Biological constrainst to farmers rice yields in three Philippine provinces. IRRI Res Pap Ser no 30
Dyck VA (1978) Economic thresholds in rice, paper presented at a short course on integrated pest control for irrigated in southeast Asia. October 16–November 18, Philippines 1978
Dyck VA, Than H, Dylay AC, Salinas GD (1977) Forecasting rice insect density and damage to plants in Asia, paper presented at the rice research station. Kerala Univ, India, December 21, 1977
Farrington, J (1972) Research based recommendations versus farmers's practices: some lessons from cotton spraying in Malawi. In: Exp Agric 13
Förster E, Rönz B (1979) Methoden der Korrelations- und Regressionsanalyse, Berl
Gomez KA (1977) On farm assessment of yield constraints: methodological problems 1–16. In: Constraints to high yields on Asian rice farms. IRRI
Gomez KA, Bernardo RC (1974) Estimation of stemborer damage in rice fields. In: J Econ Entomol 67:4
Guidelines for a surveillance and early warning system. BPI, Manila 1980, 1981
Guidelines for integrated pest control of rice insect pests (1979). FAO Plant Prot Pap no 14: 1–115, FAO ROME
Hayami Y (1978) Anatomy of a peasant economy. IRRI
Headley JC (1972a) Defining the economic threshold. In: Natl Acad Sci, Washington DC
Headley JC (1972b) Economics of pest control. In: Ann Rev Entomol, 17:273–286
Headley JC (1972c) Economics of pest control. In: Taylor RC: dynamic evaluation of pest control strategies. PhD-Thesis, Univ Columbia (unpublished)
Headley JC (1982) The economics of pest management. In: Metcalf RS, Luckman WH (eds) Introduction of insect pest management. John Wiley & Sons
Heady EO, Dillon JL (1964) Agricultural production functions. Iowa State Univ
Heinrichs EA (1978) Chemical control of brown planthopper. In: Brown planthopper threat to rice production in the tropics. IRRI
Heinrichs EA, Saxena RC, Chelliah S (1979) Development and implementation of insect pest management systems for rice in tropical Asia, I. In: ASPAC – Extension Bull no 27
Heitefuß R (1975) Pflanzenschutz: Grundlagen der Phytomedizin, Thieme, Stuttgart
Herdt RW (1979) Constraints to high yields on Asian rice farms: an interim report. IRRI
Herdt RW (1978) Some reflections an operational research for integrated pest management. In: Short course on integrated pest control for irrigated rice in south and south east Asia
Herdt RW, Jajasuriya SK (1981) The economics of insect control in rice in the Philippines. IRRI Agric Econ Dept Pap no 81-10
Herdt RW, Wickham TH (1978) Exploring the gap between potential and actual rice yields. In: Consequences of the new technology. IRRI
Hillebrandt PM (1960) The economic theory of the use of pesticides. J Agric Econom 13:52–62
Hueth D, Regev U (1974) Optimal pest management with increasing Resistance. In: Am J Agric Econom 56(3):543–552
Interagency pest control recommendations, Philippines 1980, 1981
IRRI Insecticide Evaluation Reports, 1979, 1980, 1981
Ishijura H (1969) Assessment of rice loss caused by the rice stemborer. In: The major insect pest of the rice plant. IRRI
Jahresberichte der Versuchsstationen des Bureau of Plant Industry: MRRTC, BRES, VRES, 1970–1981
Jahresberichte des Bureau of Plant Industry, Manila, verschiedene Jahrgänge
James WC (1974) Assessment of plant diseases and losses. In: Annu Rev Phytopathol, 12:27–48
Kaske R, Waibel H (1981) Ansätze für den integrierten Pflanzenschutz im Reis auf den Philippinen. In: Giessener Beiträge zur Entwicklungsforschung, Reihe 1, 7:105–125
Kelber E (1975) Möglichkeiten der Entwicklung von Befalls-Verlust-Relationen für Pflanzenkrankheiten aus Sekundärdaten. Diss Giessen
Kirby RS, Archer K (1926) Diseases of cereal and forage crops in the United States, Plant Dis Rep 53:110–208

Kisimoto R (1979) Timing of control and control threshold of the brown planthopper. In: Sogowa K, Cheng CH: Economic threshold, nature of damage and losses caused by the brown planthopper – threat to rice production in Asia. IRRI
Kranz J (1977) Die Entwicklung von Pflanzenschutzsystemen. In: Mitt Biol Bundesanst Berl 178:85–97
Kranz J (1982) Integrierter Pflanzenschutz im Reis in Südostasien. In: Entwicklung und ländlicher Raum, 16(5):22–26
Kranz J, Hau B (1981) Wie gewinnt man wirtschaftliche Schadensschwellen? In: DLG-Mitt 96:667–669
Kush GS, Chandary RC (1981) Role of resistant varieties in integrated pest management. In: ASPAC Extension Bull. no 162
Ledesma A (1982) Landless workers and rice farmers: peasant classes under agrarian reform in two Philippine villages. IRRI
Ling KC (1979) Rice virus diseases. IRRI fourth printing
Litsinger JA, Lunaban MD, Bandong JP, Pantua PC, Barrion AT, Apostol RF, Ruendi (1980) A methodology for determing insect control recommendations. In: IRRI Res Pap Ser no 46
Litsinger JA, Herrera HT, Price EC (1978) Small farmer pest control practices in three Philippine provinces; paper presented at the 9th national conference of the pest control council of the Philippines. May 3–6, 1978
Madden LV, Pennypacker SP, Antle CE, Kingsolver CH (1981) A loss model for crops. In: Phytopatol 71:7
Magallona ED (1980) Pesticide management. In: Fert Pest Authority, Philippines
MacKenzie RD, King E (1980) Developing realistic crop loss models for plant diseases. In: Teng PS, Krupa SV (eds) Crop loss assessment. Proc Stakmann EC Commemorative Symp, Dept Plant Pathology, Univ, St. Paul, Minnesota
Mumford JD (1981) Pest control decision making: sugar beet in England. Agric Econom 34:1
National food and agriculture council (NFAC) (1980) Masagana 99 plus 10, summary Manila August 1980
Norgaardt RB (1976) Integrating Economics and pest management. In: Apple L, Smith RF (eds) Integrated pest management, New York
Norton GA (1976) Analysis of decision making in crop protection. Agro-Ecosystems 3:27–44
Norton GA (1980) The role of forecasting in crop protection decision making an economic viewpoint. In: EPPO Bull 10:269–274
Ordish G (1968) Some notes on the short-term and long-term economics of pest control. PANS 14:3
Palti J (1981) Cultural practises and infectious crop diseases. In: Adv Ser Agric Sci Berl 9
Pathak MD (1977a) Defense of the rice crop against insect pests. Annu Acad Sci New York, pp 287–295
Pathak MD (1977b) Insect pest of rice IRRI Los Bañtos, Philippines
Pathak MD, Dyck VA (1973) Developing an integrated method of insect pest control. In: PANS 19:4
Peralta C, Fontilla W, Ferrer CS (1983) Economic Analysis of pest control trials. In: Plant Prot News 12:1
Philippine Statistical Yearbook (1978) Manila
Pinnstrup-Anderson et al (1976) A suggested procedure for estimating yield and production losses in crops. In: PANS 22:3
Puri S, Mullen K (1980) Applied statistics for food and agricultural scientists, Boston
Reichelderfer K, Bender F (1979) A simulative approach to the comparison of alternative methods for the control of the Mexican beatle on soybeans. In: Am Agric Econ 61:2
Reissig WH, Heinrichs EA, Antonio L, Salac MM, Santiago AC, Tenorio AM (1981) Management of insects pest of rice in farmer's fields in the Philippines. In: Prot Ecol 3
Roumasset JA (1976) Rice and risk. In: Jörgenson DW, Waelbroeck J (eds) Contrib Econ Anal 98, Amsterdam
Ruthenberg H (1977) The adoption of agricultural production services to changing circumstances: some considerations based on the experience with projects in tropical Africa. In: Agric Administration 4
Sachs L (1971) Statistische Auswertungsmethoden, Springer, Berl

Sastrosoedrjo S (1979) Agroecological aspects as related to the explosion of the brown planthopper In: Dyck VA et al. (eds) Ecology of the brown planthopper in the tropics – threat to rice production in Asia. IRRI

Satapanthy KS, Aujaneyulu A (1982) Greenhouse evaluation of emulsifiable insecticides against tungro virus infection. In: Int Rice Res Newsl 7:6

Schneeweiß H (1978) Ökonometrie, 3. Aufl, Würzburg Wien

Southwood TRE, Norton GA (1973) Economic aspects of pest management and decisions. In: Geier W, Clark LR, Anderson DR, Nix HA (eds) Insects, studies in population management, Ecological Soc. Austral; Canberra, Australia

Stern VM (1973) Economic threshold. In: Annu Rev Entomol 18:259–280

Stern VM, Smith RF, Van den Bosch R, Hagen KS (1959) The integrated control concept. Hilgardia 29:81–101

Sylvén E (1968) Threshold values in the economics of insect control in agriculture. PANS 14:3

Talpaz H, Friesbie R (1975) An advanced method for economic threshold determination: a positiv approach. In: South J Agric Econ 7:2

Tammes E (1974) Studies of yield losses In: Bradner R, Fletcher KE (eds – 1968) Insect infestation and their effect on the growth and yield of field crops. Bull Entmol Res

Van der Planck JE (1963) Plant diseases, epidemics and control. Academic Press, London

Venugopal MS, Litsinger JA (1980) Carbofuran – a direct growth stimulant of rice. In: IRRI – saturday seminar

Venugopal Rao, Subba Rao WC (1982) The importance of economic threshold in rice management. In: Int Pest Control, 24:4

Way MJ (1976) Entomology and the world food situation. Bull Entomol Soc Am 22:125–129

Weinschenck G (1965) Betriebsplanung bei unvollkommener Information. In: Agrarwirtschaft, Heft 5

Wetzel T, Freier B, Heyer W (1980) Zur Modellierung von Befalls-Schadens-Relationen wichtiger Schadinsekten des Winterweizens. In: Z angew Entomol 89:330–344

Wiese MV (1982) Crop management by comprehensive appraisal of field determining variables. In: Annu Rev Phytopatol 20:419–432

Wüsten H, Steffen G, Berg E (1981) Stand und Entwicklung des Schadschwellenkonzeptes als entscheidungsorientiertes System. In: Z Pflanzenkr Pflanzenschutz 88:465–491

Zeddies J, Waibel H (1982) Organisation und Evaluierung eines Pflanzenschutzprojektes in einem Entwicklungsland. In: 23. Jahrgang der GEWISOLA in Giessen

Subject Index

MIX
Papier aus verantwortungsvollen Quellen
Paper from responsible sources
FSC® C105338

If you have any concerns about our products,
you can contact us on
ProductSafety@springernature.com

In case Publisher is established outside the EU,
the EU authorized representative is:
Springer Nature Customer Service Center GmbH
Europaplatz 3, 69115 Heidelberg, Germany

Printed by Libri Plureos GmbH
in Hamburg, Germany